REPORTS OF DISTRICT MINING ENGINEERS

In 1919–1920 the Mining Bureau was organized into four main geographical divisions, with the field work delegated to a mining engineer in each district, working out from field offices that were established in Redding, Auburn, San Francisco and Los Angeles, respectively. This move brought the office into closer personal contact with operators, and it has many advantages over former methods of conducting field work, including lower traveling-expense bills for the Bureau's engineers. In 1923 the Redding and Auburn field offices were consolidated and moved to Sacramento.

The Redding office was reestablished in 1928, and the boundaries of each district adjusted. The counties now included in each of the four divisions and the locations of the branch offices are shown on the accompanying outline map of the State. (Frontispiece.)

Reports of mining activities and development in each division, prepared by the District Engineer, will continue to appear under the proper field division heading.

Although the petroleum industry is but little affiliated with other branches of mining, oil and gas are among the most valuable mineral products of California, and a report by the State Oil and Gas Supervisor on the current development and general conditions in the State's oil fields is included under this heading.

SACRAMENTO FIELD DIVISION

C. A. LOGAN, Mining Engineer

Mr. C. A. Logan, District Engineer, is engaged in the preparation of a special report upon the Mother Lode Gold Belt of California, which it is planned will be published soon.

SAN FRANCISCO FIELD DIVISION

C. McK. LAIZURE, Mining Engineer

Reports covering the mines and mineral resources of all of the counties in the San Francisco field division are now available, and field work at present is confined to investigations for special reports upon various economic minerals.

LOS ANGELES FIELD DIVISION

W. B. TUCKER and R. J. SAMPSON, Mining Engineers

Reports covering the mines and mineral resources of all of the counties in the Los Angeles field division are now available, and field work at present is confined to investigations for special reports upon various economic minerals.

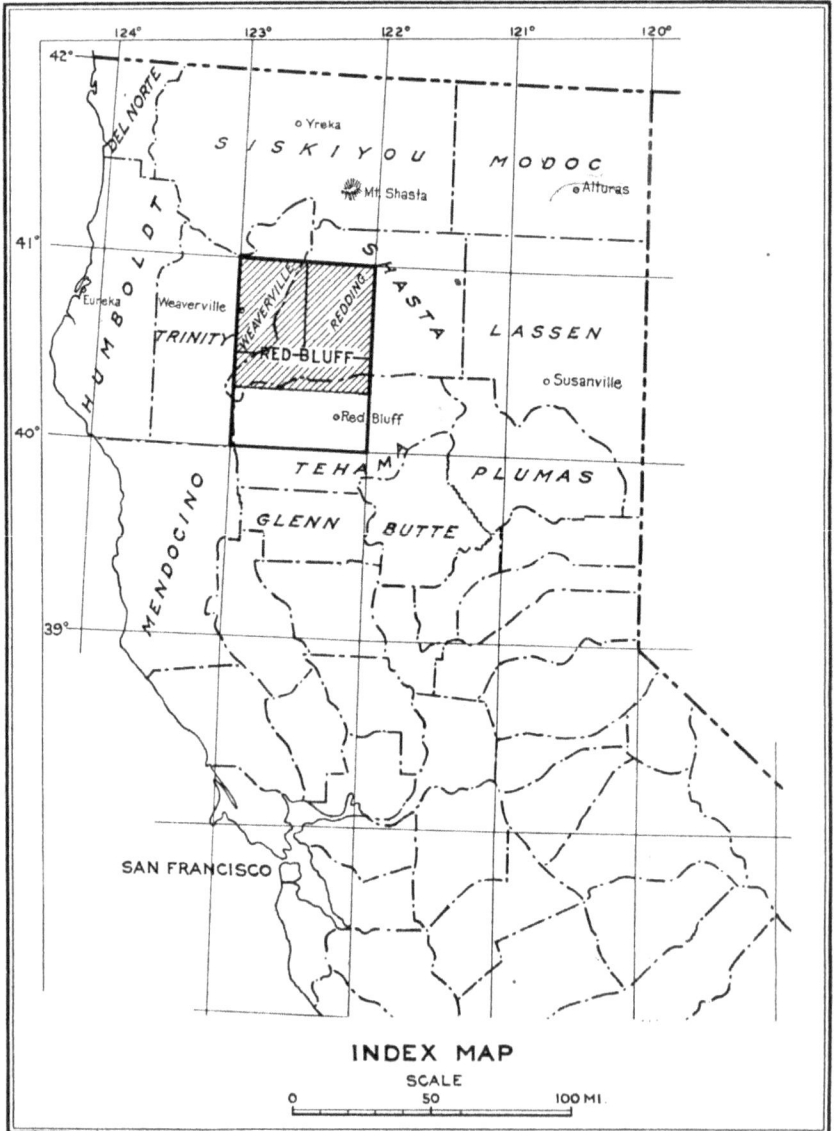

FIG. 1. Index map of northern California showing area covered in report.

REDDING FIELD DIVISION

CHAS. V. AVERILL, Mining Engineer

GOLD DEPOSITS OF THE REDDING AND WEAVERVILLE QUADRANGLES*

OUTLINE OF REPORT

Page

INTRODUCTION _____ 3

Geography _____ 3

Climate _____ 4

Industries _____ 4

Mining _____ 4

Gold Mining _____ 5

DESCRIPTION OF MINES AND PROSPECTS_____ 5

Gold (Lode)
(Mines are described in alphabetical order, numbers 1–109.)_____ 5

Gold (Placer)
(Mines are described in alphabetical order, numbers 110–135.)_____ 58

INTRODUCTION

Geography

The area covered by the present survey is that of the Redding and Weaverville quadrangles, good topographical maps of which have been published by the U. S. Geological Survey; also a small portion of the northern part of the Red Bluff quadrangle, of which only a reconnaissance topographical map is available. A degree of longitude, from 122° to 123° west, and half a degree of latitude, from 40° 30′ to 41° north, are included in the Redding and Weaverville quadrangles. They adjoin to the south the Shasta quadrangle, which was covered by a report similar to the present one in the January, 1931, chapter of State Mineralogist's Report XXVII.[1]

Redding, county seat of Shasta County, is located at the extreme northern end of the Sacramento River Valley, at the point where the river emerges from a canyon. Drainage from this central part of the area is in a general southerly direction. To the west is a divide separating Sacramento River drainage from Trinity River drainage, the latter flowing westerly to join the Klamath River near its outlet into the Pacific Ocean. Weaverville, county seat of Trinity County, is at the west edge of the area, on a tributary of Trinity River. Elevations vary

---

* This report is the economic supplement of the article following it by N. E. A. Hinds which is entitled "Geologic Formations of the Redding-Weaverville Districts, Northern California." The geologic map (in pocket) shows by a red overprint the locations of the gold mines described. The numbers used on the map correspond to those in the text.

[1] Averill, C. V., Preliminary Report on Economic Geology of the Shasta Quadrangle, State Mineralogist's Report XXVII, chapter for Jan., 1931.

from 500 ft. at Redding to nearly 9000 ft. above sea-level on the peaks in the northwest corner of the area. While lowlands suitable for farming are rather common in the southwestern part of the area, as a whole it is quite rugged and mountainous. Population is roughly 8000 or 10,000 persons.

Climate

Average annual rainfall at Redding is about 38 inches, which falls largely in the months of December, January and February, mostly as rain; although occasionally snow covers the ground for a few days at a time. At the higher elevations the amount of snow increases rapidly, and several feet are likely to lie on the ground during the winter months. The long summers are quite dry, with temperature at Redding reaching a maximum of 114° F. Weaverville is higher than Redding and has a slightly cooler climate with more snow in the winter.

Industries

Agriculture and mining are the chief basic industries, and there is some small-scale lumbering. Transportation is furnished by the main north-south line of the Southern Pacific railway, passing through Redding, and following the Sacramento River canyon. The Pacific Highway, the main north-south paved road, follows roughly the same route, but less of it is down in the canyon. A surfaced highway connects Redding and Weaverville, and a good road continues to the coast at Eureka. Another surfaced road runs easterly from Redding into the counties of Lassen and Modoc. In addition to these main roads, a large mileage of county and forest roads makes nearly all parts of the area accessible, with the exception of some of the steep mountains. Hydro-electric power is generated in large plants just to the east, and is available at reasonable rates. The resort-business, catering to vacationists, hunters and fishermen, is now attracting some attention.

The list of agricultural products is large, including fruits (but not much citrus fruit), melons, nuts, grains, hay, berries, olives, grapes, vegetables, cattle, sheep, hogs, honey and dairy products. Angora goats are raised for mohair.

Lumbering in this particular area has not reached large proportions, but big mills are operated just to the north, also at short distances to the east. Much of the area is National forest land, the part to the north being in Shasta National Forest, and the part to the west in Trinity National Forest.

Mining

The list of mineral products is also large, but for minerals other than gold the reader is referred to State Mineralogist's Report XXII.[2] The chapter for January, 1926, describes the mineral resources of Trinity County, and that for April, 1926, those of Shasta County. These publications are still available at offices of this division. They contain extensive alphabetical lists of the gold mines of the two counties, as well as descriptions of other mineral deposits. Owing to economic conditions, there is little activity at present in the development of min-

[2] Logan, C. A., Trinity Co., Jan., 1926; Shasta Co., April, 1926; Chapters of State Mineralogist's Report XXII.

erals other than gold; hence the present report is confined to deposits of that metal.

Gold Mining

Development of gold mines and prospects has been very active recently, and evidently will continue so for some time. The quartz or lode mines and prospects are described in considerable detail in the present report, and probably not very many of them have been missed during the survey. In the case of the placer mines the situation is a little different. Hundreds of persons have been at work on the streams of the area, during the past year and longer, engaged in small-scale placer mining, with pans and small sluice-boxes, into which the gravel is shoveled by hand. One result of this is that practically all public land along the streams has been taken up by placer claim locations. The gravels of the area have already been so thoroughly worked by such methods, that the average return to the present-day small-scale placer miner is barely enough to buy his food. When the small amounts produced by each individual are added up to make the statistical totals, however, the aggregate production is found to be considerable.

Little would be gained by listing these small-scale operations, and time for it has not been available. Hence only the larger and more active placer mines have been listed and described; but it is felt that enough of these have been covered to give a good idea of placer mining in the area. Of course, much gravel remains in large, unworked deposits of low grade that do not contain enough gold to pay by small-scale hand methods. It is possible that thorough prospecting of some of these deposits will reveal local areas of good grade.

GOLD (Lode)

American Mine (No. 1), owned by William Franck and H. J. Franck of French Gulch, is in Sec. 12, 13, T. 33 N., R. 7 W., four miles from French Gulch. Redding is 25 miles southeast, 18 miles to Tower House being State highway, with a good dirt road from there to French Gulch. The four miles to the mine is fair dirt road, the last mile being on a steep grade, leaving creek level in Cline Gulch at an elevation of about 1500 ft., and reaching the lowest adit of the mine at an elevation of roughly 2500 ft. Claims held are the Mill Creek Quartz Mine, Rose Gold Quartz Mine, Golden Gate Mine, Inlet Gold Quartz Mine, Alpine Extension Mine and Alpine Mine. The property is leased by C. A. Westlake, E. C. Chenoweth, and N. C. Wheeler of French Gulch. Adjoining to the east is the Gladstone mine.

An upper level shows a 2-ft. quartz vein cutting through the Bragdon formation, siliceous conglomerate and black slate. The vein shows slickensiding and occupies a former fault that cuts through the sediments at a steeper dip than that of the bedding. The strike is east and west, and the dip nearly vertical. The ore is a white quartz stained brown at this upper level by oxides of iron. Some production has recently been made from these upper levels as mentioned below. Various old workings are to be found at different elevations from the lowest, or No. 5 adit level, clear to the outcrop, some 800 or 900 ft. higher in elevation.

The No. 5 adit level runs northerly for 1150 ft. to tap the vein, on which are old drifts 150 ft. to the east and 270 ft. to the west. Stoping has been done above this level for a length of 80 ft. New work consists of a 20-ft. winze below the No. 5 level at the point where it cuts the vein. This has opened up a 2 to 3-ft. stringer zone in black slate. The ore is largely white quartz showing specks of free gold, but there is quite an admixture of black slate in lenses and horses. Some of the ore shows a banded or ribbon-structure with alternating bands of black and white. Sorting consists of removing black slate from the ore. A shipment of sorted ore, half from this winze, and half from the upper workings, of 26 tons sent to the two-stamp mill of Ernest Blagrave at the Niagara, is stated to have produced $42 per ton in free gold only. At time of visit another shipment of sorted ore entirely from the winze was sacked and piled up at the mine. This was stated to be worth more than $50 per ton in free gold.

In the old workings, lessees state that the vein reaches widths considerably greater than those given above. To the west of the new winze, at a distance of 60 ft. is an old 110-ft. winze full of water. Lessees state that at the 100-ft. level a drift was run to the west for 70 ft., and to the east 15 ft. This work was done by a company some years ago. If this winze were pumped out, some 45 ft. of drifting to the east would be required to reach a point under the new winze.

The new winze makes a little water, which is syphoned out to a point below the portal of the No. 5 adit. A rubber-diaphram pump driven by a gasoline engine is used to keep this syphon in operation. At time of visit in June, 1932, lessees were erecting a 5-stamp battery of 1000-lb. stamps near the portal of No. 5 adit. Treatment will be amalgamation followed by table-concentration.

Bibl: State Mineralogist's Reports VIII, p. 564–65; X, p. 637; XII, p. 245; XIII, pp. 349, 357; XIV, p. 778; XIX, p. 135; XX, p. 15; XXII, pp. 169, 181. U. S. G. S. Bull. 540, pp. 35, 60–61.

Amy Balch Mine (No. 2) is a property of four unpatented mining claims in the Deadwood District, Sec. 13, T. 33 N., R. 8 W. Redding is 30 miles southeast by road, 18 miles of which is State highway; the balance is the old Tom Green stage road, climbing the mountain on fairly steep grades above French Gulch. Carter states that this mine has produced $35,000 or more from veins of ochre and quartz associated with the contact of the diorite porphyry and Bragdon slate. In some places the veins cut through both of the formations mentioned. This production was made in 1913 and earlier, with hand tools, and the treatment was in arrastras. Widths varied from a few inches to a maximum of 8 ft., with an average of about 2 ft. These workings have since caved. Only assessment work was done for many years, but recently Carter has been working steadily in a lower tunnel.

This lower adit is about 100 ft. lower than the lowest of the old workings, but 250 or 300 ft. above creek level. It starts in the slate then turns and follows a slate-porphyry contact. Some crosscutting has also been done in the porphyry, the total length of the working being about 150 ft. With the exception of a seam a few inches wide, no ore has yet been found on this level. In a tunnel about 50 ft. above the creek, a drift follows a vein of crushed country rock and quartz, which is said to assay a few dollars per ton in gold for a width of 2 ft. This is in the slate within a few feet of the porphyry contact.

In view of the location of this property adjoining the Brown Bear, which has been a heavy producer, and the Lappin, another producer, and the similarity of the occurrence of the ore in all three, the Amy Balch ground appears to be a favorable place for prospecting. M. G. Carter of Redding is owner.

Bibl: State Mineralogist's Reports XIII, p. 437; XIV, p. 884.

Arps Group (No. 3) is a group of 15 patented mining claims, 212 acres, in Sec. 20, 21, 28, T. 34 N., R. 3 W., owned by William Arps of Copper City (Ydalpom P. O.), D. V. Saeltzer of Redding and A. Jaegel. Arps states that 200 tons of ore were shipped from open cuts to smelters for gold and silver. Eroded surface material pans some free gold. Old development work consisting of six tunnels and shafts aggregating over 3000 ft. in length have mostly caved. Ore found in these deeper workings contains pyrite, sphalerite, chalcopyrite, some galena, with a barite gangue. The property has not been worked recently and there is no equipment. It adjoins the Copper City group of what is commonly known as the Bully Hill mines.

Bibl: State Mineralogist's Reports XIV, p. 761; XIX, pp. 89, 90; XX, p. 427; XXII, p. 145. Bull. 50, p. 110.

Atascadero Copper Co., see Greenhorn.

Baker Mine (No. 4), two unpatented claims in Sec. 22, T. 34 N., R. 6 W., is held by J. E. Daley and J. P. Collings of French Gulch. Redding is 31 miles southeast. Redding to Tower House, 18 miles, is State highway, then 7 miles county road to a point north of French Gulch, then 4 miles of mountain road and 2 miles of trail reach the mine, which is on the East Fork of Clear Creek at an elevation of 3000 ft.

There are two adit levels with a difference in elevation of only 15 ft. The total length of the upper is 120 ft. and of the lower, 135 ft. Daley and Collings have done about 100 ft. of drifting and 30 ft. of raising between levels, also some stoping from the raise. They state that about $5,000 has been recovered from irregular pockets. A small amount of the ore has been roasted in a wood fire to free the gold from the sulphides, and it has then been amalgamated. The free gold in the pockets is recovered by panning. A hand-mortar equipped with a screen, a small amalgamating plate and a spring pole for the pestle has been used to grind some of the ore.

The ore is found in irregular masses of quartz with pockets carrying free gold, near the contact of Copley meta-andesite and Bragdon slate. Dikes of dioritic and andesitic rocks are found in the vicinity, and probably some of this is associated with the ore; but this material is so highly silicified that its original composition is uncertain. Tunnels start in Copley meta-andesite, but the ore is found mostly in the black Bragdon slate. The veins are irregular in width, in strike, and in dip. The width varies from a fraction of an inch of gouge to six and eight feet of quartz, and changes in width are abrupt.

Balaklala Mine (No. 5), consisting of 72 patented claims and 800 acres of smelter and townsite land at Coram in Sec. 10, 11, 12, 13, 14,

17, 20, 21, T. 33 N., R. 6 W., is assessed to First National Copper Co., c/o S. A. Holman, 830 Mandana Boulevard, Oakland, California. Operations of this property as a copper mine have been described in State Mineralogist's Report XXII. The mine has not been operated for a number of years, and the roads have been washed out to such an extent that the last five miles must be traveled by trail.

Considerable interest has been shown recently in the gossan deposits at this property, and the installation of a cyanide plant similar to that of the Mountain Copper Co. at Iron Mountain has been considered. The Iron Mountain plant was described in detail in State Mineralogist's Report XXVII.[3] So far as is known at the present time the Balaklala deposit is not so large as that at Iron Mountain, but such sampling as has been done indicates the possibility of a better grade. Since 1914, no work has been done on the gossan with the exception of a little prospecting done on the deposit at the Angle Station by Vickery Brothers of Kennett in the summer of 1931. The following information on this gossan is derived from a letter from S. A. Holman, who has been connected with the company since 1914, and whose address is given above.

There are three gossan deposits of some size at the Balaklala property:

(1) Gossan lying on west side of creek and slightly south and west of the Glory Hole.
(2) Gossan lying on the east side of Mule Gulch just east of the main Weil ore body.
(3) Gossan lying north of the aerial cable-tramway angle station, and known as the "Angle Station" deposit.

During the early development of this property, from 1890 to 1900, shallow shafts and tunnels were put into the gossan on all three of the above deposits. Several hundred feet of shafts and tunnels were put in the Angle Station deposit, while only a little work was done on the other two deposits. In no case was sufficient work done to make possible a close estimate of tonnage, and openings made were not sufficient to afford enough samples to show the average value. Undoubtedly good values were found in parts of the Angle Station deposit, but further work will be necessary to determine tonnage and average values. Both the Glory Hole and Mule Gulch deposits lie either directly above, or close to, known sulphide deposits, but no sulphide has yet been found in the vicinity of the Angle Station deposit. There appears to be a good chance of proving a considerable tonnage of a commercial grade of gossan at the Angle Station, while both of the other deposits have sufficient showing to warrant further exploration.

Bibl: State Mineralogist's Reports XX, pp. 427–28; XXII, pp. 145–46.

Ballou Mine (No. 6) (Manzanita) is a group of five patented claims in Sec. 7, 18, T. 31 N., R. 6 W., six miles northwest of Igo, at an elevation of 2500 ft., owned by Mrs. May V. Ballou. More than 1000 ft. of tunnels were driven on this property years ago, and some production of gold was made. Recently R. S. Ballou of Igo has leased the five patented claims, and has made two additional locations on adjoining

---
[3] Averill, C. V., The Mountain Copper Co., Ltd., Cyanide Treatment of Gossan, State Mineralogist's Report XXVII, pp. 129–138, April, 1931.

ground. Work done during the fall of 1932 consists of a 53-ft. drift adit and a 7-ft. winze on a vein not formerly worked. A shoot of pay ore from four inches to a foot wide has been found in a vein which reaches a maximum width of three feet of mineralized quartz and country rock. The ore is quartz rather heavily stained with oxides of iron, and carrying considerable sulphides, pyrite and chalcopyrite. Both walls are of quartz diorite. A shipment of 5½ tons of this ore (sorted) has recently been made to the Selby smelter. Rejects from the shipment, consisting of a few tons of mixed quartz and country rock on the dump, are stated to assay $25 per ton in gold; hence the shipment is expected to give returns considerably higher than this figure.

At the time of visit, lumber was being hauled for the construction of a new mill. Water power is available from a Pelton wheel provided with a 10-inch pipe and a head of water of 65 ft. This power is to be used to pump from a new winze that Ballou expects to sink on the vein. Additional power can be developed by using a higher ditch. The wheel formerly drove a Straub mill, now in poor condition due to the burning of the building.

Bibl: State Mineralogist's Report XXII, p. 206.

Bateman and Mauldin Prospect (No. 7) is a property of four unpatented claims recently acquired by J. C. Bateman and G. L. Mauldin of Lewiston in Sec. 14, 15, T. 33 N., R. 8 W., 45 miles from Redding, via Lewiston, by state highway and county road. The Brown Bear mine is about a mile to the east.

A quartz vein on a contact of diorite porphyry and slate is exposed in an old tunnel 225 ft. higher in elevation than the present mill-site. This varies in width from a few inches to 4 and 5 ft. A 20-ft. winze has been sunk on the wider part. The best ore is stated to be in another old tunnel near the portal, which is 125 ft. higher in elevation than the mill-site. A black gouge is found here, with in places a few inches of quartz, also lenses and spots of crushed quartz in the gouge. Widths vary from a few inches to 18 inches. Here both walls are slate.

No recent development work or mining had been done at time of visit in May, 1932. Six men were at work installing a small mill consisting of small jaw crusher about 4 by 6 inches, Ellis cannon-ball mill and Ellis amalgamator. The outfit was to be driven by 1½ and 4-hp. gasoline engines.

Beaver Mine (No. 8) (Woodfill) is a property of four unpatented claims, 80 acres, located six miles west of Ono, or 23 miles west of Redding, in Sec. 7, T. 30 N., R. 7 W., held by Geo. E. Fenby and Frank C. Sherwood of Ono. A 76-ft. crosscut taps a north-south vein, on which there is a 76-ft. drift to the south and a 26-ft. drift to the north. Dip of the vein is 80° east. Near the point where the crosscut taps the vein, a 15-ft. winze has recently been sunk, and on the hanging-wall side a streak of good ore, one foot wide, is exposed. It is not solid quartz, but is a mixture of the green schistose country rock, quartz, calcite, pyrite, minor amounts of other sulphides and possibly tellurides, and some free gold. An 800-lb. shipment of sorted ore from this streak recently sent to the Selby smelter was settled for on the basis of a gross value of $69.97 per ton. Sample assays of from $40 to $300 are said to have been obtained. The balance of the bottom of the winze, a 7-ft. width of material largely country rock, is stated by the owners to assay $9.30 per ton. The vein has been stoped out from this tunnel level to the surface. A crosscut adit, 84 ft. lower, has been driven a distance of 146 ft. Owners estimate that an additional 70 ft. will be required to reach a point under the winze.

Country rock is a dark-green serpentinized schist containing stringers of quartz and calcite. Like the rock at the Sunny Hill mine, adjoining, it has probably been derived from dioritic rocks; but it is finer in grain than that seen at the Sunny Hill.

Bell Cow Prospect (No. 9) is a property of six unpatented claims in Sec. 35, T. 30 N., R. 9 W., 18 miles southwest of Ono, on Arbuckle Mountain. S. H. Fiske of Redding is now sole owner. Workings described in State Mineralogist's Report XXII have practically all caved and are inaccessible. A surface cut, 15 ft. deep, exposes a 20-inch vein with a strike of S. 39° E. and dip 74° SW. Work recently done under the supervision of J. Ashby Simpson includes a 117-ft. adit at a point on the hillside giving a vertical depth of 99 ft. below the cut just mentioned. A winze was put down on a faulted segment of a 2-ft. vein with a very flat dip, and this is badly crushed and broken. A second tunnel at a vertical depth of 65 ft. below the one just mentioned was driven by Simpson for 200 ft. or more, and is now being continued by Fiske. Total length at time of visit was 382 ft., and no veins were exposed. Other work done by Simpson includes a 491-ft. tunnel in the gulch 200 ft. lower than the present working tunnel, and a 180-ft. tunnel 225 ft. above the working tunnel. Neither of these cut any veins. Country rock is schist of various kinds, varying from a light-colored siliceous variety to a black graphitic variety.

Bibl: State Mineralogist's Reports XVIII, pp. 296, 493–94; XXII, pp. 169–70.

Benson Mine (No. 10) (Harmont Mining Syndicate) is a group of four unpatented claims, 72 acres, in Sec. 12, 14, T. 32 N., R. 6 W., held by Chas. Harvey, 53 State St., Boston, Mass. The distance from Redding by road is nine miles, of which six miles to Shasta is State highway, the balance rough dirt road. A discovery was made here in 1922 by George Zinn. Steve Benson of Shasta bought the property in 1929, and sold to A. C. Most and Harvey the same year. The Janice, Connection, Lost Shaft lode claims are held and the S. F. placer.

Gold is found in a fissure vein with a crushed quartz and gouge filling, associated with the intrusive contact of the quartz diorite into Copley meta andesite. The vein has an average width of 3 ft., maximum of 3½ ft. in the stope. James W. Halbrook, superintendent, states that it reaches a width of 9 ft. in a 65-ft. winze, which was full of water at time of visit. A lower working driven in the summer of 1932 is stated to have drained this winze. The vein shows for a length of 15 or 20 ft. in surface cuts, and crops again at a point 125 ft. east. The dip is 70° north. A crosscut adit cuts the vein 35 ft. below where it is exposed in the open cuts, and the 60-ft. drift and the winze mentioned above are driven from this. Above the drift is a stope 25 ft.

long and 10 to 15 ft. high. Benson states that he recovered $1,150 from ore from this stope, which averaged $15 per ton in free gold.

Equipment consists of a 2-cylinder, single-stage Rix compressor, capacity 110 cu. ft. per minute at 400 r.p.m. It is belt-connected to a 10-h.p. Doak engine, which drives it at 250 r.p.m. There are also an Ingersoll jackhammer, cars, rails and drills. A 15-hp. gasoline hoist is set up at the top of a hill, 350 ft. slope distance above the mine, to pull mine cars up to the truck road. Milling equipment consists of a 2-stamp mill, 350-lb. stamps, driven by a 3½-hp. gasoline engine, and a 5-stamp mill, 1200-lb. stamps, driven by a G.M.C., S-25 gasoline tractor. Treatment is by amalgamation, and there is a Wilfley concentrating table, 5 by 10 ft. with the larger mill.

Betty May Mine (No. 11) of eight claims in Sec. 8, T. 32 N., R. 6 W., was purchased, in 1929, by Ed. Ragos of Long Beach. At that time, a 12-inch vein was developed by means of a 135-ft. incline shaft, making an angle of about 45° with the horizontal, and 100 ft. of drifting at the 100-ft. level. Ragos stated that 50 to 80 tons of $50-ore was in sight. Equipment consisted of a 10-hp. gasoline engine driving compressor and hoist; also a 3½-ft. Huntington mill driven by a 7-hp. gasoline engine. Treatment was amalgamation. Ore is a white quartz containing iron and manganese oxides and free gold. Shortly after Ragos purchased the property work was discontinued, and the mine was idle until late in 1932, when F. G. Mauthe of Los Angeles started some new development work.

Black Bear Quartz Mine (No. 12) prospect consists of the E½ of Sec. 11, T. 31 N., R. 6 W., 320 acres patented, at Mule Mountain, 12 miles west of Redding. Mrs. Mae Helene Bacon Boggs, c/o Woman's Athletic Club, 640 Sutter St., San Francisco, is the owner.

Black Spider, see Jones.

Blue Danube Prospect (No. 13) is a property of three patented claims, 45 acres, M. S. 4681, in Sec. 12, T. 32 N., R. 6 W., owned by Mrs. Clara Archer of Redding. It adjoins the Benson Mine described above. Quartz veins are found associated with quartz diorite and alaskite porphyry. An assay map furnished by Mrs. Archer indicates values from $2 to $10 per ton in gold. Width where the $10-sample was taken is 3 ft. No work has been done recently.

Bluejay Property (No. 14) is near the Paymaster, and is held by the Gifford heirs. In the summer and fall of 1932, it was reported as being opened by Morris Collins of Lewiston. On account of this work being done late in the field season, the mine was not visited.

Boswell Prospect (No. 15) is a property of some 68 unpatented mining claims in Sec. 7, T. 31 N., R. 5. W., and adjoining sections, held by Mrs. Christine G. Stevens, and leased in 1932 to F. A. Zimmerman, 1060 Court St., Redding. It is seven miles west of Redding by dirt road.

In 1929, a company called Redding Consolidated of Nevada, 315 Montgomery St., San Francisco, held this property. W. G. Campbell was president, Chas. Joseph Carey was secretary, and G. B. Reed was general manager. A 112-ft. shaft was sunk near the outcrop of a quartz vein, and this vein was apparently cut in the shaft at about the 75-ft. level, but no drifting was done. Later thousands of dollars were spent on corrugated iron buildings, steel headframe, 50-hp. electric hoist, compressor, mill containing 8-ft. by 22-inch Hardinge ballmill, No. 3 Gates gyratory crusher, amalgamating plates, Wilfley concentrating tables, Deister slimers, etc. Apparently no mining was done after the machinery was installed, and it was later sold at sheriff's sale and removed.

Several veins of white quartz outcrop on the property. At a point 300 ft. west of the 112-ft. shaft is a one-foot quartz vein with an east-west strike. On this is an old stope 20 ft. long with an average height of about 10 ft. from an adit level to the surface. The vein is on the contact of a fine-grained andesitic dike, 10 ft. wide, in the quartz diorite. From this point, and extending over 100 ft. to the northwest is a strong outcrop of white quartz with an average width of several feet. Three hundred or 400 ft. farther to the northwest is a prospect shaft exposing 6 ft. of quartz with some brown staining due to oxides of iron. A quarter of a mile to the west a parallel vein, strike northwest, dip northeast, outcrops with a width of several feet of white quartz for a length of at least 500 ft. G. B. Reed and others have claimed to have traced this vein for more than a mile, but there are probably several breaks in the outcrops in that distance; and this makes it somewhat uncertain that the same vein is being followed. This vein is associated with the dikes in the quartz diorite, similar to the one already mentioned. Several prospect cuts have been made on this vein. So far as the writer knows, no systematic sampling of these veins has yet been made to determine gold values contained.

In the summer of 1932, F. A. Zimmerman and others were installing a mill of local design on this property. It consists of a number of 12-inch iron balls rolling around in a circular grooved ring. It is similar to the Ellis mill, but is larger and contains more balls. Treatment is to be by amalgamation and table-concentration.

Broken Hills Prospect (No. 16) is a group of four unpatented claims in Sec. 23, T. 31 N., R. 6. W., 12 miles west of Redding at Mule Mountain. It is reached by dirt road and one mile of trail. H. C. Christensen and C. W. George, both of Redding, are the owners. A 400-ft. adit runs east, and cuts two north-south veins with practically vertical dip. The first shows 4 ft. of quartz with abundant pyrite and arsenopyrite, which Christensen states assays $4 per ton in gold. The second shows 22 inches of similar material, which is stated to assay $11 per ton in gold. From the portal of the tunnel, it is 150 ft. to the first vein and 100 ft. farther to the second vein. Walls are diorite, but the workings are close to the intrusive contact of this into the Copley metandesite. Breccias and other contact phases are also present.

The larger vein is found on the surface, 500 ft. south of the tunnel, and is exposed by a shallow shaft and surface cuts for several hundred feet with widths of 2 to 3 ft. About 60 ft. west of the shaft is an open cut on a 4 to 6-inch quartz stringer, from which Christensen states that $3,000 worth of ore has been taken.

Broken Hills Silver Corporation (No. 17) was reported, in 1931, to have an option on the Paymaster Mine (which see). Apparently this option was not exercised.

Brown Bear Mine (No. 18) has the largest production record of the Deadwood and French Gulch districts, which are really one district separated only by the Trinity-Shasta county line. In 1930, the property was assessed to Thomas McDonald, R. A. Foster and Ella M. Smith, c/o Thomas McDonald, French Gulch, and comprised 589 acres of patented mining claims, of which 131 acres are placer, also several hundred acres of additional patented land, in Sec. 11, 12, 13, 14, 16, 24, T. 33 N., R. 8 W. The distance from Redding is 30 miles, 18 miles State highway, the balance beyond French Gulch being a winding dirt road with fairly steep grades. A longer route, via Lewiston, avoids

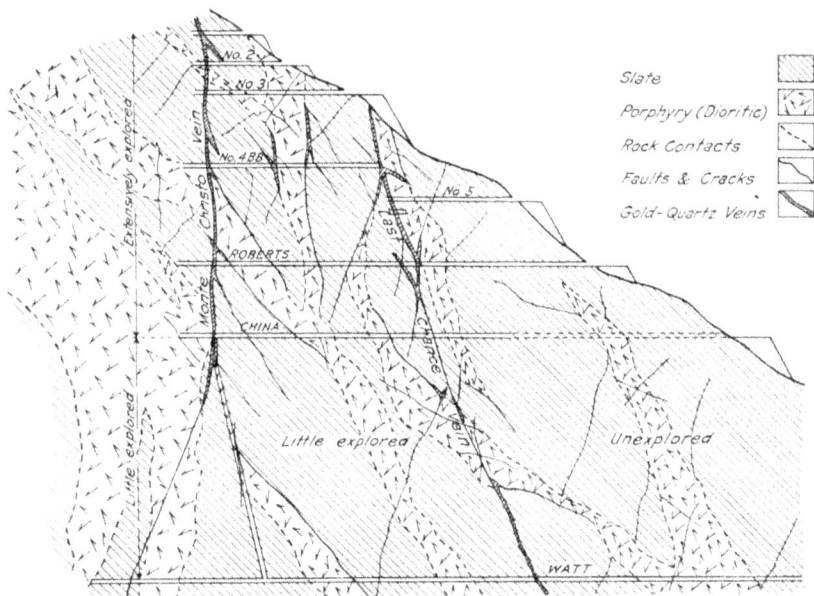

Slate
Porphyry (Dioritic)
Rock Contacts
Faults & Cracks
Gold-Quartz Veins

FIG. 2. Brown Bear Mine, generalized vertical section, looking east. Ore indicated above the China level has actually been stoped. Scale may be judged by distance from Watt level to level above—350 feet. (See description No. 18.) *Section furnished through courtesy of E. E. Erich.*

some climbing, also avoids a summit, which is at times closed by winter snows.

E. E. Erich of French Gulch is local representative of persons holding a contract to purchase the mine, and has been exploring and developing it for some time with a crew of about 15 men, in a search for new ore. He estimates [4] past gross production of the Brown Bear at $8,000,000 to $10,000,000, and states that he has definite detailed records of $3,000,000 of this. Three of its individual orebodies are stated to have produced in excess of $1,000,000 in gold each.

[4] Erich, E. E., Mining Opportunities in Known Districts, Engineering and Mining Journal, Oct. 9, 1930.

The gold occurs in veins and fissures within an area of slate, which has been cut and intruded by porphyry masses and dikes. The principal rocks are the Bragdon slates, which are found in two distinct types, with some intermediate varieties. One of these is hard, siliceous and blocky, the other is soft, black and graphtic. Intruded into these slates are the diorite-porphyry and the soda-granite-porphyry, which have been described in detail by Ferguson[5]. Extensive fissuring of all of these rocks has occured, and it is along some of these fissures that gold-bearing solutions have passed, leaving traces and concentrations of gold. Some veins can be traced 1000 to 3000 ft. They parallel and also cut the inclosing rocks at acute angles in both strike and dip. They pinch and swell abruptly. In the expansions of the veins the filling is usually quartz, and at the ends or contractions the filling becomes more an admixture of quartz with slate or porphyry or both; and finally the vein pinches down to a gouge-filled fissure. Mineralized fissures cut through both slate and porphyry, usually carrying a few inches of quartz of a high assay value; but they are often not large enough to be mined at a profit except at intersections. The ore is white quartz carrying pyrite and sphalerite, a little galena and arseno-pyrite, and free gold. The ore is usually of medium to high grade, varying from $10 to $1000 per ton. Sulphides amount to from 1% to 4% of the ore. The veins lie at varying angles but with a general east-west trend. There is a tendency to a reversal of dip on the lower levels, making the average dip nearly vertical. Principal production of the Brown Bear came from the Monte Cristo and Last Chance veins, explored 2600 ft. along their strike, and extensively mined to depths of 400 ft. Small portions of these veins have been mined to depths of 760 ft. below the surface. Several smaller veins have been explored near the surface, and have produced important amounts of high-grade ore. Both the Monte Cristo and Last Chance veins are strong fissures, with maximum widths of 20 ft; but in places they contract to a few inches of gouge. Ore was mined to a maximum width of 20 ft. and averaged between $20 and $100 per ton. The gold is not uniformly distributed throughout the veins, large portions of which contain only lean gold values.

The Watt tunnel, 350 ft. below the lowest of the old productive upper workings, had been driven 6000 ft. by previous operators into the general vicinity beneath the orebodies of the upper workings, and was then abandoned. This tunnel and some of the upper workings have been cleaned out and extended during present operations. Erich is proceeding with the idea that ore is likely to occur at intersections of fractures, particularly when one of those fractures is on the contact of the porphyry and slate. It is stated that on these contacts the slate is likely to bend when movement takes place, while the harder porphyry shatters, and open cracks result in the latter, forming a channel, which permits the passage of solutions. The structure of the slate has been studied by mapping strikes and dips, and this study indicates that folding of the slate has taken place, and that the outside crest of a fold occurs within the main workings of the mine. This is believed by the operators to be important, because stretching and shattering of the

[5] Ferguson, H. G., U. S. Geol. Survey Bull. 540-A, pp. 24-25, 1913.

slate at the outside of the fold may have opened channels favorable to the deposition of ore. In connection with this study of the mine a plan model on sheets of glass, on a scale of 100 ft. to the inch, has been constructed; also models made up of vertical sections on sheets of celluloid, on a scale of 200 ft. to the inch. A plan or section is made through the mine for each 100 ft. and these are combined to form the models.

Total length of the workings in this mine amounts to 30 miles, and time alloted to this report does not permit of a detailed description of them. Such a description would involve the reproduction of a large number of maps and sections. The mine is well equipped with air compressor, machine drills, buildings, etc. The 10-stamp mill was in operation late in 1932.

Bibl: State Mineralogist's Reports VIII, p. 639; X, p. 713; XI, pp. 483–84; XII, p. 308; XIII, pp. 440–41; XIV, pp. 885–86; XXII, pp. 15, 16, 28. U. S. G. S. Bull. 540, pp. 70–71. Eng. & Min. Jour. Vol. 130, No. 7, Oct. 9, 1930, pp. 333–34.

Buena Vista Prospect (No. 19) is a property of three unpatented mining claims in Sec. 5, T. 31 N., R. 5 W., five miles by road west of Redding. It was discovered by A. Abel about 1927, and is now held by H. G. Graves, 1732 Sixth Ave., Oakland, Cal. Art Thompson, address Redding, lives on the property.

An incline shaft has been driven on the vein, the upper 50 ft. at an angle of 30° from the horizontal, and the lower 60 ft. at an angle of 45° from the horizontal. At the 50-ft. level some faulting is evident. Drifts on this level are 38 ft. long to the north and 15 ft. to the south. At the bottom of the shaft is a drift of 40 ft. to the south. The vein averages about 5 ft. in width above the fault at the 50-ft. level, and consists of quartz and sulphides, partly oxidized, including a little chalcopyrite. Oxidation of this has produced some green staining. Below the fault, the average width of the vein is 2 ft., and the green staining is absent. Art Thompson, who did the work, thinks that this is a different vein from that in the upper part of the shaft. On the surface there is evidence that such a vein exists in the hanging wall at a distance of a few feet from the main vein. Thompson states that vein material exposed averages $30 per ton in gold. To the east of these workings, at a distance of about 1000 ft. a second quartz vein outcrops with a maximum width of 6 ft. and an average width of two feet. A 50-ft. tunnel has been driven on this. Thompson states that the assay value is $5 to $6 in gold. Equipment at the main workings consists of an 8 by 8-inch single-stage compressor belted to a 40-hp., 4-cylinder gasoline engine of marine type; also a one-cylinder, 15-hp. gasoline hoist.

Country rock is Copley meta-andesite, intruded by dikes. Just to the west of the main workings, an andesitic dike containing phenocrysts of plagioclase feldspar $\frac{1}{8}$ inch long, and smaller phenocrysts of altered biotite in a fine-grained green groundmass, was noted. What appeared to be the outcrop of a dike of rhyolitic rock of very fine grain was seen to the east of these workings.

Bully Choop Mine (No. 20) in Sec. 4, 5, 8, 9, T. 31 N., R. 8 W., adjoining the Cleveland mine on the Shasta-Trinity county-line, has until recently been assessed to Bully Choop Gold Mining & Power Co., 805 Atlanta Trust Building, Atlanta, Georgia. The patented land, about 200 acres, of this company has now been deeded to the State of California for five year's delinquent taxes. Five of the unpatented claims have been relocated by H. E. Upham and D. S. Upham of Douglas City. The elevation is 5000 ft., and steep mountain roads reach it from either side of Bully Choop Mountain. From Redding, one route is by 25 miles of county road to a point 8 miles west of Ono, then 16 miles of steep, narrow, winding, mountain road to the Cleveland mine, and a mile or two more over the summit to the Bully Choop. The other route is by 40 miles of State highway to Douglas City, then 16 miles mountain road to the Bully Choop mine.

A large amount of money was spent some years ago to equip this property (see State Mineralogist's Report XIV for 1913 and 1914). Roads were built, also dwellings, sawmill, electrical plant, 15-mile telephone line, and 30-stamp mill driven by water and electrical power. Water for power was obtained from the north and south forks of Indian Creek through a 3-mile ditch. Several thousand feet of mine workings were driven and some production resulted. Old workings are largely caved now, but the No. 2 level can be entered, being only partly caved at the portal. It is a crosscut adit reaching the Bully Choop vein 300 ft. south of the portal, and this was stoped for a length of several hundred feet to a width of 3 ft. The dip is about vertical. Roughly 300 ft. farther south, the Occidental vein is cut. This is a quartz stringer-zone in the hornblende schist, and has been stoped for a length of 200 ft., and widths of 15 to 20 ft., to the surface, roughly 150 ft. above. No. 3 level is partly caved at the portal but could be entered if a set or two of timber were replaced. According to H. E. Upham, this level reached the Bully Choop vein, and considerable work was done on it; but it did not reach the Occidental, and about 50 ft. of work will be required to reach the latter. No. 7 level (caved), 1000 ft. lower in elevation, is stated not to have reached either of the veins mentioned, but considerable stoping was done on a third vein. The old 30-stamp mill of 1050-lb. stamps is still on the property, also a 4½-ft. Huntington mill, rails, cars, etc.

At the time of visit in July, 1932, Upham brothers had been working for two weeks on their locations. Work was stated to be on the Occidental claim. Several small open cuts had been made on stringers near one end of the stope where the Occidental vein was worked to the surface. The greatest depth reached was about 10 ft. below the surface. The stringers are about vertical and vary in width from a few inches to two feet. The best ore is in a white quartz heavily stained with yellow and some red ochre; and vugs and cavities are filled with these iron oxides from the oxidation of the original sulphides. This ore pans well in free gold. Occasionally specks of free gold can be seen in the white quarts away from oxides of iron. Plans called for the immediate milling of some of this ore in the Huntington mill, with treatment by amalgamation.

Country rock is hornblende schist, which is not very uniform in composition here, and appears to be the result of pressure acting on a variety of igneous rocks, with some recrystallization. Original rocks were probably mostly diorites and gabbros, but in one wall of the Occidental vein is a rock that apparently was a dike of 'bird's eye porphyry'. It contains white remnants of feldspar crystals, a quarter of an inch long, in a groundmass of brown micaceous material, which shows the schistose structure.

Bibl: State Mineralogist's Report VIII, p. 640; XII, p. 308; XIII, p. 441; XIV, p. 886; XX, p. 182.

California Progressive Mining Co., see Clipper and Snyder.

Calmich Mine (No. 21) (Lappin) is a group of four patented claims assessed to Mary F. Bacon and others, 60 S. San Rafael Ave., Pasadena, California. It is in the Deadwood District in Sec. 13, T. 33 N., R. 8 W., at an elevation of 3500 ft., and is reached by means of about a mile of steep mountain road starting at the lowest mill-level of the Brown Bear mine. Harvey Clayton and H. E. Kimmel of Lewiston are leasing the property. Clayton states that former operators extracted a shoot 90 ft. long, with 200 ft. of backs, from a vein averaging 6 inches wide, and with a maximum width of 10 inches. Ore was treated by amalgamation, and the recovery is stated to have been $60 per ton. Another irregular shoot with a maximum length of 30 ft. was extracted. The strike is east and west, dip north, and the shoot raked to the east. Clayton and Kimmel have sunk a 40-ft. shaft from an upper tunnel level on an average incline of 60° from the horizontal, and have driven a 15-ft. crosscut from the bottom. Within the next 10 ft. they expected to hole into a place into the old workings, where they stated that some good ore was exposed. The shaft and crosscut are in hard slate. A two-stamp mill and 4½-hp. gasoline engine have been installed. Treatment is amalgamation.

Calumet Mine (No. 22) is an old patented property of three claims in Sec. 10, T. 32 N., R. 5 W., assessed to Mrs. Mary B. Garlick, 2243 Franklin St., San Francisco. W. D. Tillotson of Redding is local representative. The ground covers extensions to the south of quartz veins found at the Walker property in the Old Diggings District. Quartz outcrops with widths of 4 ft. and occasional swells to double that width can be seen on the property. Workings are largely caved. No mining has been done here for many years.

Bibl: State Mineralogist's Reports VIII, p. 563; X, pp. 631–32; XI, pp. 43, 395; XIV, p. 781.

Central Mine (No. 23) is a group of four patented claims and a millsite in Sec. 3, 4, T. 32 N., R. 5 W., and 33, 34, T. 33 N., R. 5 W., owned by A. A. Anthony and others of Redding. It is 11 miles by road north of Redding, and is just across the Sacramento River from the main north-south line of the Southern Pacific railroad.

Workings of this property are caved, and can not now be entered; but it is of considerable interest as a possible member of a large group in the Old Diggings district. Possibilities of grouping the properties of this old district are discussed further under the heading 'Old Diggings.' Outcrops of some of the veins can be observed on the surface. The following description of the vein-system and workings is quoted from Logan[6]:

"There are four veins on the property, three of which strike north and dip 65° east and the fourth strikes east. Work has been confined to the north-striking Central vein, which has been mined through adits. The lowest of these is the only one now accessible. It is a crosscut to the west for 956 ft., where it cut the

[6] Logan, C. A., State Mineralogist's Report XXII, pp. 170–171, 1926.

vein and thence followed it north about 650 ft. and also followed it southward until it pinched. This level gave about 335 ft. of backs, and known ore above it has been stoped. One stope in the north drift was 150 ft. long, 30 ft. high and 10 to 15 ft. wide. There was six feet of solid quartz on the footwall, the balance of vein being a stringer lead. The country rock is altered andesite. On the south, about 400 ft. of work failed to develop ore. On this side, the vein pinched down to nothing.

"This adit was run from about the center of Central Extension claim, through the Central claim to within 400 ft. of the north end-line. About 300 ft. west, on the Shasta claim, is the parallel Shasta vein, and the Pocket vein, also striking north, lies 250 to 300 ft. east of the Central. The east-striking vein crosses the north end of claims, 300 to 400 ft. north of the present face of Central adit, depending on dip of east-striking vein, which at the surface is about 45° north.

"In the earlier work between 1885 and 1895 Huntington mills were used for crushing ore, the millsite being on the river some distance from the mine. It was found that only about 50% of the gold was recovered by this method. Later, shipments to Selby smelter, Keswick smelter and finally to Kennett were made. The total output is estimated at about $500,000, but smelter receipts for much of this have been lost. Receipts in Anthony's possession for shipments to the Keswick smelter between 1900 and 1906 show assay values ranging from $3.40 to $84 a ton, little of the ore going below $6 a ton, and the average for six years being over $13 a ton. Of course, much lower grade ore was shipped later, as it could be cheaply worked and delivered, and the quartz was worth $2 a ton as flux alone. A tramway delivered ore from the mine to railroad.

"The lower part of this mine could be prospected to advantage from the Reid Mine shaft, which is nearby. There is no equipment at the Central."

Bibl: State Mineralogist's Reports VIII, p. 565; X, p. 631; XII, p. 246; XIII, p. 351; XIV, p. 782; XVIII, p. 494; XXII, pp. 170–71.

Chanchelulla Prospect (No. 24) is a group of eight unpatented claims and a millsite in Sec. 1, T. 30 N., R. 10 W., held by David Boyer and F. W. Forrester, both of Ono, Shasta Co., Cal., and D. L. Martin, c/o Harry Glover, Redding. It is 60 miles by road west of Redding, county road to Harrison Gulch or Knob (P.O.), then narrow, winding mountain road with fairly steep grades to Deer Lick or Combs Springs, then one mile steep mountain trail to mine, which is at elevations of from 3000 to 5500 ft. The millsite is on a fork of Browns Creek, which flows a good stream.

A 3- to 4-ft. quartz vein, exposed in open cuts and one 50-ft. adit, strikes S. 50° E. and dips 47° SW. Forrester states that this has been traced for the length of four claims, 6000 ft.; but it is doubtful that the vein could be opened with widths as great as indicated above for this entire distance. Cuts to a total number of 10 expose the vein at various points for the length of these four claims. Depths vary from one foot to 10 ft. The 50-ft. tunnel reaches a depth of about 50 ft. on the vein. A second vein, one foot wide, called the Shaft Vein, is opened by means of an incline shaft, 17 ft. deep. The strike is S. 30° E., dip 30° SW. This shaft is 75 or 100 ft. northeast of a point on the main vein, and hence intersections on both the strike and dip are indicated. Forrester states that this vein assays $16 to $20 per ton in gold and a few ounces in silver. Both walls of the Shaft Vein are gabbro-schist or gneiss, and the hanging wall of the main vein is of this formation also. The footwall is a breccia, white in color, and probably rhyolitic. It contains quartz fragments of a maximum size of three-quarters of an inch in diameter.

Cleveland Mine (No. 25), on the Shasta-Trinity county line in Sec. 9, 16, T. 31 N., R. 8 W., is the property of the C. F. Foster Co., of Corning. It has been optioned for some time to C. J. Kerr and associates of Redding and Eugene, Oregon. In area it totals 800 acres of patented ground, all of Sec. 16, and patented mining claims, Gen. Washington, Ida M., etc., to make up the balance. By road it is

42 miles west of Redding, 25 miles of county road to a point eight miles west of Ono, 17 miles of steep, narrow, winding mountain road to mine, which is at an elevation of 5500 ft.

A quartz fissure vein carrying plentiful sulphides of iron at shallow depth, some arsenic and a little copper, has been mined for gold. A little free milling ore was found at the outcrops. Following along the outcrops for 1000 ft. or more, one sees pinches and swells in the vein, which varies in width from a few feet to 20 ft., average 8 or 10 ft. The strike is N. 9° E. and the dip 60° E. Exposures in outcrops and tunnels in this upper part indicate a length of about 2500 ft. C. J. Kerr states that a vein corresponding in strike, and in the right position for the dip, is found on Jerusalem Creek, 2000 ft. lower in elevation. If this is the same vein, a length of 3500 ft. is indicated. Country rock is a complex schist series, mica-schist and diorite- or gabbro-gneiss. One small lens of limestone was noted.

No. 1 tunnel starts on the outcrop, and is 140 ft long. From this a winze 60 ft. deep has been sunk, and some stoping has been done on a width of 6 ft. of quartz. No. 2 tunnel, 150 ft. lower and 400 ft. long, is caved. No. 3 tunnel, 50 ft. below No. 2, is 425 ft. long, part cross-cut and part drift, from which the vein has been stoped out. The vein can be seen, continuing to the south at the south end of the stope, with a width of about 4 ft. of quartz, with no drifting on it. To the north, in the face of the tunnel, the vein is disturbed by faulting; but it can probably be picked up again with a small amount of work. A winze, 75 ft. deep, from this level is full of water. No. 4 tunnel, 258 ft. below No. 3, is 820 ft. long, and is entirely in country rock. It was started to tap the vein at a point under the winze from No. 3, but according to present operators it must be driven 200 ft. farther to reach that objective. At time of visit, work had been confined to cleaning old workings. An old 10-stamp mill operated here years ago is in a wrecked condition.

Bibl: State Mineralogist's Reports XII, p. 310; XIII, p. 442; XIV, p. 887.

Clipper and Snyder Group (No. 26) is an old group of patented claims assessed to California Progressive Mining Co., 126 Folsom St., San Francisco, and located in Sec. 36, T. 34 N., R. 6 W. It is reached by eight miles of trail starting at the Squaw Creek bridge about a mile west of Kennett. When the Uncle Sam mine was visited, in the summer of 1932, short stretches of this trail were in poor condition, so that a horse could not get over them. Workings of the Clipper and Snyder are caved, and the property was not visited during the present survey. The following information is from a report by Melville Atwood, dated Sept. 21, 1914, and handed to the writer by C. R. Hussey, who states that he has controlling interest in the corporation. The Snyder vein is stated to be 3 to 4 ft. and more in width. On this were a No. 1 level, 125 ft. long, a No. 2 level, 310 ft. long, exposing the vein with widths of from 2 to 6 ft., and a No. 3 level, 360 ft. long exposing similar widths. The yield is stated to be from $5 to $9 per ton including sulphurets and specimen-ore. A run of 415 tons in the ten-stamp mill of 850-lb. stamps is stated to have yielded in the mill $4183, and in specimens shipped to the smelter $12,300. The Lillian Maude vein is stated to show one to one and a half feet of $20-ore. The Carrey Huston vein is stated to be 2 to 6 ft. wide and more, with a 525-ft. tunnel on it. Ore from the entire length of this is said to have averaged $9 per ton.

The Clipper vein was developed with a 326-ft. tunnel, and 140 ft. from the portal, a 44-ft. winze, which exposed a vein 6 ft. wide for the entire distance. Ore from this winze is stated to have assayed $10 per ton, and from the top of the tunnel $8 to $30 per ton. Ore is stated to contain 1½ to 2% of sulphides assaying $150 to $300 per ton.

Bibl: State Mineralogist's Reports X, pp. 640–41; XI, p. 399; XII, p. 246; XIV, p. 781.

Craig, see Mason and **Thayer.**

Crown Point and Midnight (No. 27) are two patented claims in Sec. 13, 24, T. 31 N., R. 6 W., assessed to C. H. Hammond, W. 328-8th Ave., Spokane, Washington. They are ten miles by county road southwest of Redding, the last mile of road not being passable by auto at present. A quartz vein with widths of 4 and 5 ft. strikes northwest and dips 50° to the northeast. A 150-crosscut adit cuts the vein at a point which gives backs to the surface of 100 ft. From the crosscut there is 50 ft. of drifting and a raise to the surface. A small amount of ore was stoped here and shipped to the Keswick smelter years ago, but the results are not known. The ore is a white quartz containing pyrite. On the hanging wall is a streak of six inches or more of dark green gouge, which also carries sulphides of iron. H. C. Christensen recently concentrated some of these sulphides by panning, and had the concentrates assayed. He states that the return was $22 per ton in gold. The workings described are on the Midnight claim. To the northwest, at a distance of 1000 ft., just across the property line on ground held by Geo. P. W. Jensen, 320 Market St., San Francisco, and leased to H. C. Christensen of Redding, is an incline shaft on the same vein, open for 40 ft., but full of water below. To the south, on the Crown Point claim is a shaft on the vein, now caved. The vein may be seen in a road-cut nearby. Christensen has recently cleaned out a 150-ft. tunnel here, and started on the shaft.

A peculiar dike was observed on this ground, apparently a coarse-grained igneous rock with quartz crystals from a quarter of an inch to a half inch in diameter. This appears to have been a dike in the Copley meta-andesite, and to have been subjected to great pressure with the meta-andesite, so that it now resembles some of the older schists of the region. Some of it has abundant muscovite on the cleavage planes. The Copley meta-andesite has been subjected to such great pressure here that much of it resembles slate, except for the green color.

Democrat Group (No. 28) contains 15 unpatented claims, 280 acres in Sec. 12, 13, T. 33 N., R. 10 W., 1½ miles from Weaverville on the ridge between West Weaver Creek and Democrat Gulch, held by R. H. Junkans, C. J. Hanna and D. E. Ryan of Weaverville. J. W. Shaw has recently cleaned out a 340-ft. tunnel, cutting sandstone, then a serpentinized dike, then an andesitic dike. The last 50 ft. is in schist, most of which is light green and chloritic appearing, but there are layers of graphitic schist. The strike is N. 35° E., and the dip 45° NW in the face, but considerably flatter than that in other parts of the tunnel. The schist carries lenses and veins of bluish quartz showing sulphides of iron, which is stated to have been very productive of gold on parts of the surface and in the placers nearby.

On the opposite side of the ridge (west fork of Weaver Creek side), at a point 1600 ft. east of the tunnel mentioned above, and 150 ft. lower in elevation, is a second old tunnel, 160 ft. long, in the schist. At the face is a width of 14 ft. of quartz heavily mineralized with sulphides of iron. Samples were being taken of this at time of visit, but the results were not learned.

Bibl: State Mineralogist's Report XXII, pp. 17, 29.

Dixie Queen (No. 29) is a group of three unpatented claims and a fraction held by Melvin A. Taylor, 809 Eighth Avenue, Oakland. It is located in Sec. 14, 23, T. 32 N., R. 9 W., on Spring Gulch, a tributary of Indian Creek, reached via Douglas City. A road has been started to go up Spring Gulch from Indian Creek, but it is not yet complete. This property is stated to have produced $5,000 worth of a base shipping ore, containing gold, silver, lead and zinc, worth $200 per ton and over, from a shoot 120 ft. long and 40 ft. high to the surface. At time of visit, John de Cordi was working on this same vein on an 8-inch width of ore, said to assay $30 per ton. Total length of working adit was 300 ft. Equipment consisted of an arrastre operated by water power with an 80-ft. head, used in a one-inch nozzle. Treatment was by plate-amalgamation.

Bibl: State Mineralogist's Reports XIV, p. 887; XXII, pp. 17, 29.

Double Header Mine (No. 30) is an 80-acre tract of patented land bought from the railroad in Sec. 33, T. 33 N., R. 7 W., and is assessed to Chas. Garrett of Schilling and Wm. P. Donnelly of Anderson. It is on the Redding-Weaverville State highway, 20 miles west of Redding. Some production was made here in 1888, at which time a 5-stamp mill was operated. The ore was produced from surface workings on the contact between slate and decomposed Copley meta-andesite. At a point 150 ft. below these old workings, a tunnel was driven by the old company, and the surface near the portal produced some good ore; but most of the 400 ft. driven at that time was a low-grade vein with both walls of Copley meta-andesite. The strike is a little east of north, dip 45° east, and the width varies from a few inches to 5 ft. of quartz. In 1931, Garrett let a contract for driving this drift-tunnel ahead for 120 ft. on the vein, in the hope of reaching a part with a slate hanging wall. Walls are both Copley meta-andesite in the new part of the tunnel, and at time of visit the vein had not yet been sampled. The 2-stamp mill, with 6-hp. gasoline engine, has not been used recently.

Bibl: State Mineralogist's Reports VIII, p. 567–68; XVIII, pp. 138, 354, 405; XIX, p. 11.

Eastman Con. Mines Corp.—see Venecia.

East View (Winnie) (No. 31) comprises three unpatented claims in Sec. 6, T. 32 N., R. 6 W., held by Chester Vergnes of Oak Bottom. It is just north of Oak Bottom, which is 12 miles west of Redding on the Weaverville highway. A quartz fissure vein, which varies in width from one inch to two feet, is developed by means of a 500-ft. adit, of which 100 ft. is on the vein, giving a depth of 50 ft. below the surface. It is stated to contain good values in gold. From the opposite side of the hill to the workings mentioned above, a crosscut adit has been driven for a distance of 40 ft. Country rock is Copley meta-andesite cut by quartz-diorite dikes.

Eiller Mine (No. 32) consists of four unpatented mining claims in Sec. 11, T. 32 N., R 6 W., held by H. M. Huse, O. E. Stone and Dr. H. E. MacDonald of Redding. From Redding it is 10 miles by road; the first 6 miles to Shasta is State highway; the balance of 4 miles is rough dirt road; and the last mile is very steep.

When Dr. Bell and brothers held this property in 1880 to 1889, a 400-ft. adit was driven; and about 125 ft. lower in elevation, a second adit thought to be 1900 ft. long was driven. At that time the mine was equipped with a mill of two 1000-lb. stamps and a canvas plant, which was probably used very little. In 1928–1929, the Gold Hill Mining Co., with A. C. Marr in charge, cleaned out and retimbered 700 ft. of this lower tunnel. According to H. M. Huse, at this 700-ft. point the tunnel exposes a width of 11 ft. of quartz that assays $1.00 per ton in gold. He thinks that the grade improves beyond this point. At the present time, a small cut exposes a quartz vein, 3 ft. wide, heavily mineralized with oxides of iron, from which good pannings in gold can be obtained. The strike is about east and west and the dip vertical. At a point 75 ft. lower in elevation than the cut, or at the same elevation as the 400-ft. adit (caved) mentioned above, and at a distance of 100 ft. from the latter, a new adit was being driven at time of visit by O. E. Stone and two other men. The course of the adit was southwest for 42 ft.; it then turned to the right to follow 6 to 8 inches of gouge with occasional bunches of quartz, which was thought to represent the 3-ft. vein at this level. Drifting was being continued to the west to reach a point under the surface-cut, 50 or 75 ft. ahead. Country rock is quartz diorite, and there is a rhyolitic dike 200 ft. southwest of the workings. Equipment consists of a cook-house and a bunk-house to accommodate 10 men, and a blacksmith shop.

Ellis (No. 33) is a single patented claim of 15.8 acres in Sec. 6, T. 33 N., R. 5 W., assessed to L. C. Monahan of Berkeley and M. E. Dittmar of Dunsmuir. According to Dittmar, the long adit starting in the Bonanza claim of the Uncle Sam group, and running northeasterly, cut a vein in the Ellis claim showing 18 to 20 inches of quartz and chalcopyrite that assays $8 in gold and silver, and 2½% copper. This adit was run by the American Zinc, Lead and Smelting Co. of St. Louis in a search for extensions of ore bodies in the Mammoth mine. The face is at a point 700 ft. southeast of the northwest corner of T. 33 N., R. 5 W., M. D. M. (a township corner). Backs on this vein from the point where cut to the surface are stated to be over 1300 ft.

Bibl: (American Zinc, Lead and Smelting Co.) State Mineralogist's Report XIX, p. 57.

Evening Star (No. 34). A claim of this name, 18 acres, is held by Wm. Blagrave of French Gulch. It is between holdings of the Washington and Milkmaid in the French Gulch District.

Evening Star (No. 35) a group of this name will be found shown on the map included herewith under the heading 'Old Diggings District.' No work has been done on the group recently but it is interesting as a possible member of a large group that could be formed in the district mentioned. It was formerly held by Geo. Bayha and W. D. Tillotson of Redding, and is patented ground.

Bibl: State Mineralogist's Report XIV, p. 786.

Evening Star (No. 36) formerly Bright Star or Iron Mask, is a property of three claims in Sec. 30, 31, T. 33 N., R. 6 W., held by W. H. Reel of Schilling, and Mrs. Mabel Barndt, Ivanhoe Hotel, Berkeley. It is just west of the Mad Mule.

A crosscut adit cuts the vein at a distance from the portal of 190 ft. The vein strikes about north and south, has a steep dip to the east, and a width of from one to $3\frac{1}{2}$ ft. of quartz. There is 600 ft. of drifting on this level, and a pay-shoot 60 ft. long. A winze was started on this pay-shoot; but at a depth of 30 ft. it passed out of the shoot, which has a rake to the north. There is a level at 125 ft. with drifts to the north of 130 ft., and to the south of 100 ft. At a distance of 60 ft. from the winze in the north drift, a raise was put up; and this struck the pay-shoot. W. H. Reel states that he did this work, and that the face of the raise showed specimen ore, none of which was stoped. The width of the vein was 18 inches, and it appeared to be widening to the north in the single drift-round that was put in. There is a cave on the main tunnel level at the collar of this winze, and none of this lower work can be examined at the present time. Above the main level, the pay-shoot is stoped out. The vein is on the contact of slate with alaskite on the main level for a short distance, the balance having both walls of alaskite. A parallel vein, similar in general appearance at a distance of 25 or 30 ft. has been exposed on this level, and can be seen in a short drift. Nothing has been done in these workings for years, assessment work being above in an open cut on a small stringer.

Bibl: State Mineralogist's Reports XII, p. 250; XIII, p. 359; XIV, p. 780. U. S. G. S. Bull. 540, p. 55.

Fairview Mine (No. 37) comprises 520 acres of patented mining claims and 330 acres of unpatented claims, including quartz claims, placer claims and mill sites, in Sec. 3, 10, 11, T. 34 N., R. 8 W., assessed to Fairview Mining Co., Room 1406, 315 Montgomery St., San Francisco. It is on the Trinity River 10 miles north of Lewiston, but from the road along the river it is accessible by foot-bridge only. A steep mountain road connects the mine with Trinity Mountain Ranger Station, north of French Gulch. The mine has been worked at various times for many years, in 1901 to 1908 by the company, later by different lessees.

At time of visit, in June, 1932, lessees were Don Matheson and Max Jacobs of Lewiston, and J. Mackey of French Gulch. At an elevation 650 ft. higher than the lowest adit level (No. 4), they were driving an adit to cut a faulted segment of the vein, that had been developed by surface cuts just to the east of the point where the east end of the old stopes came to the surface. The adit had been driven 100 ft. by the three men working on three different shifts, at the rate of 8 ft. per day, with hand drilling. Sixty feet more was expected to be sufficient to reach the segment of the vein, with 50 or 60 ft. of backs. From the open cuts above, ore had recently been mined; and it was treated in a 3-ft. Huntington mill equipped with amalgamating plates. At the portal of No. 4 adit, is a 90-cu. ft. compressor driven by a one-cylinder gasoline engine. Of the old 40-stamp mill, 10 stamps still remain on the property, but they are not in working order.

The vein strikes in a general easterly and westerly direction, and dips to the north at angles of 60° and greater. It has been stoped out for a vertical height of 450 ft. from a shoot having a flat rake to the east. Stopes are several hundred feet long, and average about six feet in width. The No. 4 level, the lowest that has been driven, is 200 ft. below the lowest stoping, and several hundred feet higher than Trinity River. It is the only level now open, and is 1700 ft. in length. Lessees have recently cleaned out an old 30-ft. raise near the face, raised an additional 70 ft., drifted 65 ft. east from the top of this raise, and raised another 30 ft. from the face of the drift. This work has exposed some vein material dipping south, or opposite to the normal dip of the vein.

FIG. 3.   Fairview Mine.   Tailings from Lewiston Dredge in foreground.
(See description No. 37.)

Some of it has been of very good grade, but the vein material has been very irregular and difficult to follow. The width is about 2 ft.

Both walls of the vein are black slate. A fault-contact of this with the meta-andesite is exposed on the No. 4 level, but no ore has been found on such a contact. The ore recently found in the raise above No. 4 level appears to have associated with it a dike rock. This is very dark in color, perhaps due to an admixture of black slate. Crystals of free quartz indicate that it may have originally been rhyolitic. Alteration and silicification make its composition obscure. Some of it has a brecciated appearance.

Bibl: State Mineralogist's Report XIV, p. 888. U. S. G. S. Bull. 540, pp. 75–76.

Fields-Meckel Prospect (No. 38) an unpatented group on the south slope of Weaver Bally, at an elevation of about 6000 ft., is held by Dr. D. B. Fields of Weaverville and A. C. Meckel. The property has been visited by the writer, but caving at the portal of the main adit prevented examination of the workings. Logan [7] has described these as follows:

The lowest adit is at 5810 ft. elevation and follows the vein N. 75° E. for a distance of 230 ft. The vein is from six inches to two feet wide and forms small lenses of solid white quartz, the longest of which is 35 ft. long. There is a raise 32 ft. above the adit. The vein is tightly frozen to both walls, which are hard hornblende schist. The best showing so far on the property was at the portal of this adit, where $2500 is said to have been extracted from a hole two feet deep on the vein. The vein shows six inches of quartz at the face, and is said to pan well in free gold as far as drifted. It shows free gold in places. There are three other adits, the lowest, now caved, being 50 ft. below the one described, and two short ones recently started 100 ft. and 150 ft. above the main adit. All these are in an area of a few acres and the rest of the claims are unprospected except for shallow pits, although the vein is said to be traceable for 2000 ft. Assessment work only is being done, and there is no equipment except hand tools.

Bibl: State Mineralogist's Report XXII, pp. 19-20, 29.

First National Copper Co., see Balaklala.

Fisk, Jim, see under 'J.'

Five Pines Mine (No. 39) comprises 80 acres of patented land and five unpatented mining claims in Sec. 19, 20, T. 35 N., R. 7 W., 40 miles by road north of Redding, owned by Van Ness Brothers of French Gulch. The mine has been a good producer at intervals since its discovery in 1896 by H. J. Van Ness. The gold is found in pockets associated with a slate-andesite contact. Lester Van Ness estimates the total production at $350,000 to $400,000, and states that the largest pocket yielded $44,000. The ore in these pockets is very rich, and it is estimated that 60% of the total production has been made from ore that was pounded out in a hand-mortar.

The slate-andesite contact strikes southeast and dips 45° to the northeast. The main vein strikes east and west, and has a very steep dip to the north. This is a former fault fracture, which displaces the slate-andesite contact about 20 ft. The vein averages $4 to $7 per ton where both walls are andesite, but a rich shoot with a flat rake is found where the vein cuts the slate-andesite contact. In addition to this main vein there are a number of parallel stringers which also form rich pockets at the intersection with the contact mentioned. The andesite is very fine-grained gray igneous rock, with individual crystals too small to be identified without a microscope. The slate is a soft, black variety, that contains a large amount of graphite.

On the Surprise claim is a 450-ft. drift adit, which reaches the collar of an inside shaft or winze. This was started in the slate, 32 ft. from the contact, out in the hanging wall; but the contact flattened, and the lower part of the winze is in andesite. There is more than 1000 ft. of other work on this main level, partly caved. James Skeen, of Redding, deepened the incline winze, in 1926, from the 200 to the

[7] Logan, C. A., State Mineralogist's Report XXII, pp. 19-20, 1926.

300-ft. level, and did 125 ft. of drifting on a stringer, which was stoped; and some high-grade was taken out. The main vein has not been developed below the 100-ft. level. It will require 300 or 400 ft. of work to reach this vein on the 300-ft. level. Equipment here consists of an Ingersoll-Rand 4-drill compressor, a large Sullivan portable compressor, three Cameron sinking pumps, the largest being No. 6, and complete hoisting equipment on the winze. No work has been done here recently, and this equipment is not in use. Water stands in the winze to within a short distance of the main adit level. A small mill consisting of crusher, ball mill and amalgamating plates stands here also, but it was not a success in treating the ore. A two-stamp mill stands on another part of the property.

On the Emma claim two men are working above a 300-ft. adit near creek level. East-west veins cutting the slate-andesite contact are being explored by means of various raises and drifts above the adit. Free gold associated with calcite and arsenopyrite is found in pockets near this contact, and some production is being made from ore crushed in a hand-mortar. Rich pockets come at the intersections of stringers with the slate, especially if the stringers have an abrupt change in direction near such an intersection. Three men are working in a narrow channel below the mine, but on the same property, for placer gold. This channel extends for at least three-quarters of a mile below the workings on the Emma claim, and is 35 ft. wide and has an 8-ft. depth of gravel, which is stated to average 50c per cubic yard. The gold is partly quite coarse, a $3-piece and an $8-piece having been recovered recently. Lester Van Ness says that in earlier years, one 14-ounce piece and several one-ounce pieces were recovered.

Bibl: State Mineralogist's Reports XIV, p. 889; XVIII, p. 355; XXII, pp. 20, 29. Pre. Rep. No. 8, p. 18. U. S. G. S. Bull. 540, pp. 73-74.

Florence Claim, see Phillips.

Franklin, see Milkmaid.

Gambrinus (No. 40). One claim of this old mine in Sec. 16, T. 32 N., R. 6 W., is being held by W. E. Saunders of Schilling, who is living on the property. Tunnels and other underground workings are caved, and a few surface cuts are all that can now be examined. The mine was discovered in 1870, and according to Ferguson[1], in 1912, a known production of $127,000 had been made; and some additional production had been made, of which no record had been kept.

Bibl: State Mineralogist's Report XIV, p. 786. U. S. G. S. Bull. 540, pp. 38, 39, 50, 51.

Ganim Gold Mines Company (No. 41), J. S. Ganim of Redding, president, holds a group of 14 unpatented claims in Sec. 5, 8, T. 32 N., R. 6 W., 2½ miles northwest of Schilling. The lowest adit level on this property has been extended to a length of 900 feet, with branches additional as mentioned below. It exposes a body of talc with a maximum width of about 60 feet between two bodies of siliceous material carrying

[1] Ferguson, H. G., Gold Lodes of the Weaverville Quadrangle, U. S. G. S. Bull. 540-A, pp. 44-45, 1913.

sulphides. Most of the sulphide is pyrite, but there is some chalco-pyrite, sphalerite, and possibly a little galena. These bodies of siliceous material have about the same width as the talc, and are stated by Ganim to assay from $1.50 to $2 in gold with an occasional sample as high as $50. The strike is northwest and southeast and the dip 45° to the northeast. A west drift from this main adit is 250 ft. long, and there are two crosscuts from the latter with a total length of 70 ft. Four raises in the talc average 50 ft. each. From a second drift, 110 ft. long, a raise of 140 ft. reaches the surface. There are also a 70–ft. winze and crosscuts of 100 ft. and 30 ft. from this drift. Cuts here and there on the surface along the strike of the talc indicate that it has a possible length of 1000 ft. or more. The last talc shipped was in 1929, 12 to 15 carloads bringing $10 to $10.50 per ton, f.o.b. San Francisco. Freight and hauling costs $4.50 per ton.

FIG. 4.   Kennerly gold saving machine, tried for short time at Ganim mine.
(See description No. 41.)

At time of visit, Ganim was working on a quartz vein higher on the hill. It is a vertical vein showing 3 to 4 ft. of white quartz containing various copper minerals, sulphides, carbonates, etc.; and is stated to assay $8 to $10 per ton in gold. The crosscut adit is 100 ft. long, and drifting from it amounts to 100 ft. If the main level is extended 500 or 600 ft., it will cut this vertical vein, giving backs of 435 ft. Equipment consists of Chicago Pneumatic compressor, 9 by 14 inches, 325 cubic ft. per minute, driven by a semi-Diesel 50-hp. engine, drill sharpener, air drills and air hoist.

Bibl: State Mineralogist's Reports XVIII, pp. 730-31; XIX, p. 11; XX, p. 15; XXII, pp. 171-72, 210-11.

Garfield, see Lone Cedar.

Gibson, Molly, see under 'M.'

Gladiator Gold Mining Co. (No. 42) (Hiatt) owns a patented claim, the Cleveland, M. S. 5197, in the NE¼ Sec. 15, T. 32 N., R. 5

W., about five miles by road north of Redding. According to an old map by Albion S. Howe, on file at the Redding office of this division, the mine was developed by a shaft on the vein to the 200-ft. level. On the 100-ft. level, about 250 ft. of drifting was done, and two shoots of ore, each about 100 ft. long were stoped. On the 200-ft. level, only 25 or 30 ft. of drifting was done. The vein varied from a foot to three feet in width, averaging about 14 inches, and the dip was very steep. The collar of the old shaft and the top of a raise, 30 ft. east of it, are both caved at the present time (1932) and are choked with old timber a short distance below the surface. Where the vein is exposed in these openings, it is a stringer zone, with quartz stringers one or two inches in width in andesite. Total width of this stringer zone is given above. There is no equipment of any value.

Gladstone Mine (No. 43) in Sec. 1, 7, 8, 12, 18, T. 33 N., R. 6 & 7 W., 4½ miles northeast of French Gulch is the property of Hazel Gold Mining Co., C. F. Kimball, secretary, 1103 First National Bank Building, San Francisco. This is the deepest mine of the region, and has a production variously estimated at between $3,000,000 and $5,000,000. The vein occurs in the slate, sandstone and conglomerate of the Bragdon formation, in a crushed zone 60 ft. wide. The upper part of the mine to a depth of 1000 ft. was worked through adits, from the lowest of which a shaft was sunk 1400 ft., with levels about 100 ft. apart. The mine has been idle for more than four years, and the lower part is full of water. Considerable equipment, including a 30-stamp mill of 1050-lb. stamps and good buildings, remains on it. Further details may be found in the references given below, of which the April, 1926, chapter of State Mineralogist's Report XXII is available at offices of the Division of Mines at 25c per copy.

Bibl: State Mineralogist's Reports VIII, pp. 568–69; X, p. 637; XI, p. 45; XII, pp. 248–49; XIII, p. 357; XIV, p. 787; XVIII. pp. 43, 96, 256; XIX, p. 11; XXII, pp. 172–73. U. S. G. S. Bull. 540, pp. 35, 37, 46, 57–60.

Globe Mine (No. 44) in Sec. 15, 16, 21, T. 35 N., R. 10 W., four miles north of Dedrick, at an elevation of 6100 to 6700 ft., is an old mine on which a 20-stamp mill and cyanide plant were operated in 1913, and later. It has been idle for a number of years; workings have caved; and equipment has been taken away. Details concerning the old workings will be found in the January, 1926, chapter of State Mineralogist's Report XXII, copies of which are available at offices of the Division of Mines at 25c per copy.

Bibl: State Mineralogist's Reports X, pp. 711–12; XI, p. 483; XII, pp. 309–10; XIII, pp. 432, 442, 447; XIV, pp. 889–91; XVII, p. 540; XXII, pp. 20, 21. U. S. G. S. Bull. 540, pp. 76–78.

Gold Hill Mining Co. see Eiller.

Gold Leaf (No. 45) or Crown Deep is an old property of 122 acres, unpatented, in Sec. 5, T. 31 N., R. 5 W., recently transferred by E. P. Preble to D. F. Densel and W. S. Standish of Los Angeles. A small amount of cleaning out of old workings is all that has been done recently.

Bibl: State Mineralogist's Reports VIII, p. 358; XVII, pp. 521–22; XVIII, pp. 296, 298, 354, 406, 495; XIX, p. 11.

Golinsky Mine (No. 46) in Sec. 28, T. 34 N., R. 5 W., is assessed to Golinsky Copper Co, c/o W. D. Tillotson, Redding; and is leased to H. M. Vickery and W. C. Vickery of Kennett. From the electric substation near Kennett, 2½ miles of steep one-way road follow the old electric railway grade to the Holt and Gregg quarry. From here it is about a mile and a half by trail to the mine, which is located on the steep side of the canyon of Little Backbone Creek. Vickery Bros. leased the mine with the idea of treating the gossan that overlies the copper ores formerly mined here, for gold. A plant using plate-amalgamation was first constructed, but this treatment was not successful, and it was decided to convert it to use the cyanide process, from which good results have been obtained on this type of ore by the Mountain Copper Co., at Iron Mountain. A heavy storm destroyed some of the buildings and cyanide equipment in the winter 1931–32, and the plant has not yet been completed. It was designed to treat 200 tons per day. A sample map of the Golinsky, on file in the Redding office of this division, indicates that the gossan here carries very good values in both gold and silver; but the tonnage is much smaller than that at Iron Mountain.

Bibl: State Mineralogist's Reports XIV, p. 766; XX, p. 432; XXII, p. 149; Bull. 50, p. 100.

Great Western Gold Mines, Ltd. (No. 47), a Washington corporation, is purchasing from H. C. Christensen, Knob Route, Redding, four unpatented mining claims, in Sec. 22, T. 31 N., R. 6 W., in the Muletown district, through A. O. Witte of Seattle. Fiscal agents are Comonett Co., 1411 4th Ave., Seattle. The property is 12 miles southwest of Redding by dirt road.

On this group, a gulch cuts through a vein; and drift-adits have been run both ways from it. The drift to the northeast follows a vein with vertical dip, and width from a few inches to three feet, average say 18 inches, possibly 2 ft. where a small stope has been made. Christensen says that 40 tons of $16 ore were shipped from here to the Keswick smelter years ago. The length of the northeast drift is 350 ft., and two crosscuts have been run near the face to a practically parallel vein, 15 ft. to the northwest, and 50 ft. of drifting has been done on this second vein. It shows a few inches of quartz and a 2-inch streak of nearly solid pyrite in one crosscut. The other drift-adit, running 150 ft. to the southwest, follows the vein for a short distance, and it then pinches down to a width of only a few inches. It has been traced for 500 or 600 ft. farther in this direction by means of surface-cuts and shallow shafts, and is thought by the operators to be the same vein on which a small body of good ore has recently been discovered at the Potosi mine, half a mile to the southwest. Christensen states that at a point 300 ft. southwest of the portals of the two adits he has recently extracted 10 tons of ore that free-milled $11 per ton. Additional values in the sulphides were not saved. The vein is associated with a fine-grained greenish-gray dike rock (probably dacitic) in the quartz diorite.

Machinery from a small mill, erected on this property about 1928, has been removed with the exception of a Size O Gates gyratory crusher, belted to a 3-hp. gasoline engine. At the time of visit, in

July, 1932, Great Western Gold Mines, Ltd., had done no new work
on the claims.

Greenhorn Mine (No. 48) comprising 15 patented mining claims
and 120 acres of other patented land in Sec. 6, T. 32 N., R. 7 W., 22
miles by State highway northwest of Redding, is assessed to Atascadero Copper Co., c/o Frank H. Proctor, 14th Floor Tribune Tower,
Oakland.

Copper ore occurs here as lenticular bodies of heavy sulphide in a
shear-zone in altered rhyolite, which strikes N. 60° W., and dips 40°
SW. This sulphide ore is chiefly chalcopyrite associated with pyrite
and pyrrhotite. The management claims about 250,000 tons of 3%
copper ore blocked out. There is also a certain amount of secondary
copper ore of very high grade, containing chalcocite, cuprite, native
copper, and a little malachite and azurite. When the price of copper
was around 18c per pound in 1929 and early 1930, a number of car-
loads of this ore were shipped to a smelter at Tacoma. The grade was
high enough so that a profit was made in spite of the long truck and
railroad haul. At the same time development work was done in a
body of gossan that overlies these ores, and samples were taken which
showed very good values in gold. However, what tonnage of this
gossan was developed is not known to the writer. The mine has been
idle for about two years, and efforts to obtain further data from the
owners have not so far been successful. Albert Hanford, 791 High-
land, Piedmont, California, is interested in the company, and was in
charge of the last work done there. There are buildings on the property but no machinery.

Bibl: State Mineralogist's Reports XVII, p. 518; XX, p. 433;
XXII, p. 149.

Green, Tom, see under 'T.'

Halcyon Mine (No. 49) comprises five unpatented quartz claims,
Halcyon, Ruby Pearl, Midway, Globe and Long Pine, in Sec. 17,
T. 33 N., R. 7 W., held by C. C. Fox and Wm. C. Blagrave of French
Gulch. It is on the steep mountain, adjoining the Washington mine,
and the distance from Redding is 25 miles. The Halcyon Syndicate,
J. H. Scott, President, Merchants Exchange Building, San Francisco,
was operating the property late in 1932. Under the supervision of A.
A. Wren of French Gulch, 15 to 18 tons per day of a soft, decomposed
surface ore were being put through a 3½-ft. Huntington mill, and
treated on amalgamating plates. Operators stated that the recovery
by this method was high, and amounted to $8 to $12 per ton. The
crew consisted of eight men. Power was furnished by the Pacific Gas
and Electric Co.

The ore is a soft, decomposed diorite-porphyry, found near its
intrusive contact with the Bragdon slates. The diorite-porphyry contains numerous quartz stringers, an inch and less in width; and the
more of these that are present, the better the ore is stated to be. A few
larger stringers are found also, stated to contain high-grade ore. The
diorite-porphyry is so decomposed that no blasting is necessary in mining it, only picking. Several hundred feet of development work have

been done recently at a maximum depth below the surface of 50 or 75 ft., and it is the ore from this work that has been milled.

Bibl: State Mineralogist's Report XIV, p. 789.

Happy Jack (No. 50) is a group of six patented claims in Sec. 28, 33, T. 32 N., R. 6 W., 10 miles west of Redding, recently acquired at a tax sale by H. H. Shuffleton, Jr., of Redding. The property is stated to contain several quartz veins that carry gold, with widths as great as 8 ft., in the reference below. The Mineral Survey number is 3163.

Bibl: State Mineralogist's Report XVII, p. 522.

Harmont Mining Syndicate, see Benson.

Harrison Gulch (No. 51) (Midas, Victor and Twinvict Groups). The location is in Sec. 3, 4, 10, T. 29 N., R. 10 W., in the southwestern corner of Shasta Co., 52 miles from Redding. J. B. Moore of Knob, Shasta Co., is the present owner, having acquired the property at a tax sale. Very extensive mining operations were carried on here for years, but nothing has been done recently; and much of the equipment has been removed. For a description of former operations see April, 1926, chapter of State Mineralogist's Report XXII, pp. 173–74, copies of which are on sale at offices of this division at 25¢ each.

Bibl: State Mineralogist's Reports XIV, p. 792; XXII, pp. 173–74; Pre. Rep. No. 8, p. 18.

Hiatt, see Gladiator.

Hoboe (No. 52) is an unpatented claim in Sec. 17, T. 33 N., R. 7 W,. in the French Gulch District, near the Washington and Niagara mines. It is held by Ernest Blagrave of French Gulch, who has made some recent production from a 2-ft. quartz vein with a vertical dip. A lower crosscut adit was being driven to tap this in July, 1932. The ore is treated in a 2-stamp mill owned by Blagrave and located near the Niagara mill. Small lots of custom ore from other parts of the district are crushed here also. The only treatment is amalgamation.

Houston, Sam, see under 'S'.

Independence (No. 53) is a group of seven claims in Sec. 4, 32, T. 29 & 30 N., R. 10 W., in the Harrison Gulch district, held by John Kutz of Knob. The owner states that a number of quartz veins have been developed by different workings on the group with widths as great as 6 ft., and good values in gold. Two hundred tons have been milled. This property was not visited during the present survey, due to lack of time.

Iron Mountain, see Mountain Copper Co.

Isabel and Queen (No. 54) are two unpatented claims in Sec. 8, T. 32 N., R. 6 W., in the Whiskytown district, held by F. C. Darrow of Schilling. He is prospecting these alone, and has two open cuts about 10 ft. deep on either side of a small gulch. They expose a shear-zone in greenstone, striking S. 60° E., and with a steep dip to the northeast. The width is about 10 ft. There are a few quartz stringers from a fraction of an inch to an inch wide in this shear-zone, and the mixture of this with the country rock (Copley meta-andesite) pans a little free gold. Darrow states that he recently took out a $40-pocket about 200 ft. away, but not on this same shear-zone. Several hundred feet of old tunnels are open on another part of the claims, and quartz showing plentiful chalcopyrite is piled on the dump. This quartz is stated to have come from an old winze, now full of water.

Jensen Group (No. 55) is in the Centerville District, in Sec. 24, T. 31 N., R. 6 W., and adjoining sections, and is held by Geo. P. W. Jensen, 320 Market St., San Francisco. The group is now leased to H. C. Christensen, Knob Route, Redding, who states that Geo. Morrison and Wm. Hague produced $2400 here a year ago, from a 125-ft. incline now full of water. Small seams of gouge carrying lenses of pyrite a few inches wide are seen in prospect cuts.

Jim Fisk (No. 56) Santa Claus and Orifice are three patented quartz claims in the Muletown District, Sec. 15, T. 31 N., R. 6 W., assessed to Mrs. Frances Klapetsky of Long Beach, California. In the reference given below is described a mill of odd design of 2½ tons capacity in 24 hours, which was operated here for a short time about 1893. All that can be seen now are a few caved, shallow shafts and open cuts on a quartz vein, 2 ft. wide, that follows a dike (probably rhyolitic) in quartz diorite. No mining has been done here recently.

Bibl: State Mineralogist's Report XII, p. 251.

Jones Property (No. 57) is in Sec. 19, T. 32 N., R. 5 W., at the old smelter town of Keswick, six miles by road from Redding. It comprises 44 acres of patented land bought from the railroad and one unpatented mining claim, held by E. F. Jones and Rosa Jones, Box 371, Redding. The old Black Spider mine is included.

Quartz veins from a few inches to 2½ ft. in width, occasionally 4 ft., are found here, associated with dikes in the quartz diorite. The general appearance of these dikes is that of a fine-grained basic rock, but in one place an acidic appearance was noted, possibly due to secondary silicification. Jones states that the bunches of high-grade ore found here have been at the intersections of the quartz veins and the dikes, and that the largest of these bunches removed by him produced $1200 from a few tons of ore.

In 1931, P. T. Atwood and A. C. Most did some work on the property. A shaft was sunk 80 ft. and 40 ft. of drifting was done near the bottom. Water now stands in the shaft to within 35 ft. of the surface. An adit, which penetrated some of the old workings, was opened up for a distance of 120 ft., requiring spiling and heavy timbering. Besides the hoisting machinery, a very light stamp mill, depending on springs to add force to the drop of the stamps, was installed. All of this machinery has since been removed, and Jones is working the property alone.

Bibl: (Black Spider) State Mineralogist's Report XII, p. 246; XIII, p. 350; XIV, p. 779.

Jumbo Prospect (No. 58) comprises 40 acres of patented land in Sec. 18, T. 31 N., R. 5 W., six miles southwest of Redding. It was formerly worked for placer gold, and shows a bedrock of Copley meta-andesite intruded by dikes of quartz diorite; also quite a few quartz stringers. The owners, Ira Smith and others of Redding, think that the whole mass contains enough gold to make mining on a large scale profitable. Whether this is true, or whether the gold is confined to the quartz stringers and fractures in the bedrock into which fine particles of placer gold have found their way, can be determined only by development work. Old superficial development work consisting of open cuts and shallow shafts has caved.

Bibl: State Mineralogist's Report XVII, p. 522.

Kanaka Prospect (No. 59) (Sunshine) is a group of four unpatented claims in Sec. 28, T. 32 N., R. 6 W., nine miles by road west of Redding, held by John R. Kemble of Schilling. Quartz veins, carrying gold in pyrite and chalcopyrite, are found associated with the intrusive contact of the quartz diorite into Copley meta-andesite. On the Dorothy claim is a crosscut adit, 80 ft. long, from which 200 ft. of drifting has been done on the vein. At time of visit, about half of this drifting was not accessible on account of caving. Average width of the vein is 14 inches, maximum 5 ft. The strike is east and west, and the dip 60°. At the intersection of the crosscut adit with the drifts is a winze full of water, depth unknown. Several parallel veins are known, and, judging from dumps, there are several hundred feet of workings on these, now caved. A white quartz vein with an average width of 4 ft. and a maximum of 10 ft. outcrops for a distance of 500 ft. on the K. N. H. and Bullfrog claims; but there is no development work on it. A small Huntington mill was placed on this property in 1928–1929, but it was later removed; and no production resulted.

La Clair Prospect (No. 60) comprises 20 acres of patented land bought from the railroad, and assessed to L. E. Van Ness, 4920 Park Boulevard, Oakland; also two unpatented mining claims held by Robert Mack of French Gulch. The patented land is in Sec. 13, T. 35 N., R. 8 W., and the claims in the adjoining Sec. 24. The distance from Redding is 42 miles, 18 miles state highway to Tower House, 23 miles dirt road to confluence of Van Ness Creek and Trinity River, boat across river, and one mile steep trail to the claims, 1000 ft. above river level. They are just across the river from the mouth of Van Ness Creek, on which the Five Pines mine is located. A flat incline shaft follows the slate-andesite contact down for 60 ft. The strike is east and west and the dip 20° north. Drifts near the bottom of the incline expose the contact for 12 ft. in one direction and 14 ft. in the other. At the bottom, a 12-ft. raise from a 350-ft. adit level below connects with the incline. On the contact are bunches and lenses of quartz about 14 inches in width that pan free gold. Mack estimates the value at $12 to $15 per ton. On the 350-ft. adit level is another drift on the contact, 25 ft. long, showing irregular bunches of quartz. Of the 350 ft., total length of this main level, 50 ft. near the face is drifting on a second vein entirely in the black slate. It shows 4 to 5 ft. of mixed quartz and black slate, which Mack states assays $2.50 per ton in gold. A 10-ft. raise from the adit level exposes this vein, with a dip of 45° to the west. The property was idle in the summer of 1932, but assessment on the unpatented claims was to be done later.

Lappin, see Calmich.

Lila (No. 61) is a claim in the Deadwood District just above the Brown Bear Mine, assessed to T. E. Thurner and Gruss estate, 1130 Chestnut St., Chico. It is in Sec. 12, T. 33 N., R. 8 W. J. B. Grigsby, address Lila Mine, via French Gulch, is working the property on lease and option. At time of visit two men were moving thirty tons of sacked ore extracted in former operations so that it could be loaded on trucks and hauled to the small stamp mill on the Lappin or Calmich property. Prospecting was to be done later in the old workings consisting of several hundred feet of tunnels with some stoping on quartz fissure veins carrying free gold and pyrite. Country rock is typical of French Gulch and Deadwood districts, Bragdon slates intruded by porphyry dikes. Veins observed in the workings are steeply dipping fissures in the hard dioritic porphyry.

Little Alice, see R. A. M.

Lone Cedar Prospect (No. 62) (Garfield) comprises two unpatented claims in Sec. 34, T. 33 N., R. 5 W., ten miles north of Redding, held by Wm. Goll, Buckeye Route, Redding. The last mile is trail at present, but an old road could be graded out to the portal of the working tunnel at small expense. A surface cut exposes a quartz vein with a width of 14 to 18 in., showing oxide of iron and some spots of chalcopyrite. At time of visit in the spring of 1932, Goll was cleaning out the portal of an old tunnel below, which he thinks is 140 ft. long, and in which he states that the vein is exposed with a width of $3\frac{1}{2}$ ft. A hundred feet or more lower in elevation is a crosscut adit, several hundred feet in length, which is caved at the portal. It probably has not yet reached the vein. Equipment consists of car and track. Country rock is a rhyolitic dike intruded into Copley meta-andesite. The northern properties of the Old Diggings District are just to the west on the other side of a low ridge.

Bibl: State Mineralogist's Report XIV, p. 787.

Los Andes Gold Mining Co. According to C. E. Bass of Tacoma, a company of this name is to take over the Clipper and Snyder mines, (which see) equip and operate them immediately. It is a Nevada corporation.

Lucky Boy Prospect (No. 63) is a group of seven unpatented mining claims in Sec. 15, T. 33 N., R. 5 W., held by Dr. H. C. Erno of Redding, and Clay Vance of Portland. It is two miles south of Kennett by a rough road, and is practically on the Sacramento River, only the right-of-way of the Southern Pacific being between the property and the river. Part of it was formerly called the Red Cut, named from a red bank facing the river.

Numerous quartz stringers from an inch to a few inches in width, and with various strikes and dips are found in the alaskite porphyry here. Some 20 pockets are said to have been mined from these stringers by means of shallow cuts and shafts, with a total gross production of $20,000. At time of visit, in April, 1932, one man was employed, drifting on a 4-inch quartz stringer in the face of a 200-ft. adit. Present owners are holding the property with the idea of trying to develop a large body of low-grade ore that can be mined with power shovels.

Three samples taken by Alex M. Wilson of Oroville are stated to have given gold assays from $1.25 to $4 per ton.

Lyons Mine (No. 64) is a group of four fractional patented quartz claims 48 acres in Sec. 22, T. 33 N., R. 5 W., adjoining the National. It is the property of James P. Bradner and others, 977 Mission St., San Francisco. Some stoping was done here on a quartz vein, and the ore was removed through an adit-level, years ago.

Mad Dog Prospect (No. 65) includes three unpatented claims held by L. E. Alpaugh and C. A. Williams of Redding in Sec. 10, T. 32 N., R. 6 W., 18 miles northwest of Redding. The last three quarters of a mile is steep mountain trail. A sheer zone about 45 ft. wide is found in a fine-grained silicified alaskite, and in this zone are quartz veins from a few inches to 3 ft. in width, average say one foot. This quartz pans a little very finely divided free gold. Development consists of one 146 ft. adit level, which was driven by means of picking only, as blasting is likely to start a cave. According to Alpaugh, some good gold assays were obtained from grab samples taken from the broken material removed in driving the tunnel.

Maddox Mining Co. (No. 66) (Mad Ox) (Caribou Gold Mining & Power Co.). This mine is in Sec. 28, T. 33 N., R. 6 W., and is assessed to Maddox Mining Co., c/o Grace Schilling, Redding. Three claims were purchased by the company, and additional mining claims have been located. W. H. Grant, Box 64, Palo Alto, is president, and E. E. Erich of French Gulch is engineer. By road, the mine is 16 miles northwest of Redding, 11 miles state highway, balance mountain road. It was idle when visited, and underground workings were not entered. The following description of these was furnished by W. H. Reel of Schilling, who owned an interest in the property. He states that total work on the main adit level amounts to 2000 ft., 900 ft. of cross-cutting and 1100 ft. of drifting, and that two shoots of ore are exposed, each 60 ft. long and 2 to 4 ft. in width. These are 100 ft. apart. At a distance of 600 ft. is a shoot of lower grade, 50 ft. long, maximum width 20 ft., value $6 to $8 per ton. A winze, 135 ft. deep, from this level was full of water. Reel says that there is 100 ft. of drifting from the bottom of this, showing a 6-ft. vein but no pay shoot, and that on the 90-ft. level is a drift 135 ft. long.

Equipment includes an Ingersoll portable type compressor with a capacity of two drills, driven by gasoline engine; also Leyner drill-sharpener. 10-stamp mill of 850-lb. stamps, jaw crusher and amalgamating plates.

Bibl: U. S. G. S. Bull. 540-A, p. 45.

Mad Mule Mine (No. 67) (Banghart) is a patented property of 43 acres in Sec. 31, T. 33 N., R. 6 W., five miles due north of Whisky-town by mountain road. Whiskytown (Schilling P. O.) is on the Redding-Weaverville highway. The holding, Lot 37, has been divided into separate parcels, one assessed to Geo. Jackson, R. 1, Box 169, Los Gatos, the other to J. W. Curry and others, 589—31st Street, Oakland. M. T. Thomsen, 1835 Santa Clara Ave., Alameda, is stated to have an interest in the latter parcel. The mine is a typical pocket mine, free gold being found in small shoots, but at times in single pieces worth several thousand dollars each. It is associated with an intrusive dike of striking appearance, containing feldspar crystals, **a** half inch long, in a fine-grained gray groundmass, in which are minute hornblende needles. The dike cuts the meta-andesite, the alaskite, and the conglomerate and slate of the Bragdon formation; and its width is about 150 ft. The free gold is usually found in the slate at its contact with the dike, and at 'points,' where there is a sharp change in direction, like the crest of a fold, the bottom of a trough, or an intersection with a fault-plane. According to W. H. Reel of Schilling, who has recently been leasing here, workings amounting to about 8 miles of drifting and raising on the contact have been driven in the search for such pockets. Total production is roughly estimated at over $1,000,000.

Bibl: State Mineralogist's Reports IX, p. 38; XI, p. 397; XII, p. 252; XIII, p. 361; XIV, p. 791; XIX, p. 11; XXII, p. 175. U. S. G. S. Bull. 540, pp. 40, 42, 52–54.

Mammoth, see old Diggings; also U. S. Smelting, Refining and Min. Exp. Co.

Mason and Thayer (No. 68) (Craig) is a property of 17 claims in Sec. 33 (approx.) T. 35 N., R. 10 W. The survey to subdivide this area into sections has not yet been made. J. B. and F. J. McCauley, 524 Gough St., San Francisco, are the owners. From Redding, it is 75 miles to the northwest, 60 miles State highway through Weaverville, balance mountain road, steep, narrow, and winding, the last mile being for tractor only. Quartz veins carrying gold and sulphides are found in hornblende schist; and dioritic dikes, six to 60 ft. wide are associated with them. Workings consist of a mill-tunnel level of a total length of 2000 ft., 1200 ft. of which is drifting. On a level 70 ft. higher is 260 ft. of drifting, and there are two raises between these levels. On a level 60 ft. still higher is 260 ft. of drifting and a raise to the surface. According to Keith MacKay, superintendent, this work blocks out 9000 to 10,000 tons of $6 ore in a shoot 170 ft. long and with a width of 3 ft. Strike is N. 70° E., and dip 45° S. In a tunnel below the mill-tunnel level, 350 ft. of drifting has recently been done on a body of base ore that assays $3 to $5 per ton.

Equipment includes numerous buildings, a new 100-hp. hydro-electric plant on Canyon Creek, 173-cu. ft. Chicago Pneumatic compressor with 45-hp. Hercules gasoline engine, drill sharpener, 22-hp. Cletrac tractor, and a sawmill of 5000 board feet capacity in 8 hours. There is also a 5-stamp mill, 6 by 5 ft. ball mill, crusher 14 by 12 inches, two flotation machines, two Deveraux agitators, concentrating tables, and electric motors. Only a few hundred tons of ore have been milled with this equipment, and it was idle at time of visit. Six men were employed on development work.

Bibl: State Mineralogist's Reports XIV, p. 887; XXII, p. 16. U. S. G. S. Bull. 540, p. 78.

Meeks Property (No. 69) consists of patented land in the SE¼ of the NE¼ of Sec. 15, T. 32 N., R. 5 W., about five miles by road north of Redding. A small quartz vein was developed some years ago by means of a vertical shaft. The collar of the shaft is now caved. An old five-stamp mill stands on the property. The vein is in the contact zone between the alaskite and the Copley meta-andesite.

Menzel Mine (No. 70) comprising the Scottish Chief and Santa Clara patented claims, in Sec. 31, T. 33 N., R. 5 W., in the Flat Creek Mining District, is owned by Wm. Menzel of Redding. Old workings on the surface, consisting of cuts, tunnels and an 80-ft. shaft, have caved. Some production was made from these about 1880 or 1890. By means of outcrops and these old workings, the vein can be traced for at least 1000 ft. on the surface. Widths vary from a fracture containing no quartz to a maximum of 6 ft. of quartz, with some country rock between strands of the vein. The strike is S. 85° W., and the dip 60° or 70° to the south. A fine-grained, basic, pre-mineral dike, one or 2 ft. wide, accompanies the vein.

A main tunnel level, roughly 150 ft. below the outcrop, crosscuts in the hanging wall of the vein for 600 ft., then turns and follows the vein to the west for 200 ft. The width varies from nothing to a maximum of 8 ft. of quartz at the face. This face is stated to sample $4 to $5 per ton in gold. Near it is a raise, and 35 ft. above is an intermediate level with 140 ft. of drifting to the west and 25 ft to the east. The width here varies from nothing to 6 ft., with an average of 3 ft. Above this intermediate, at the raise, a stope has been carried up 30 ft. high and 30 ft. long, and a raise from the west end of this has tapped some of the old caved workings at a height of 20 ft. above the face of the stope. Chas. Hammond, who has recently cleaned out all of these old workings and has done some additional raising, removed 30 tons of sorted ore from the east end of this stope, which he states returned $18 per ton by amalgamation. Ore in the face of the stope is a white ribbon quartz, and a little chalcopyrite was seen in it. According to a report on file in the Redding office of this division, written by J. E. Miller, the vein continues into the adjoining Murrey claim of The Mountain Copper Co. Hammond's equipment, consisting of $3\frac{1}{2}$-ft. Huntington mill and amalgamating plates has since been removed. He ground the ore to pass a 40-mesh screen.

Bibl: State Mineralogist's Report XIV, p. 797.

Midas, see Harrison Gulch.

Milkmaid and Franklin Mines (No. 71) have been idle for a number of years, and have not been visited by the writer. A description of them will be found in the April, 1926, chapter of State Mineralogist's Report XXII, pp. 175–76, copies of which are available at offices of this division at 25¢ each. There is also a bibliography in the same report p. 183.

Minnesota Mine (No. 72) includes 34 acres patented, M. S. 3585, and one unpatented claim in Sec. 1, 2, T. 32 N., R. 6 W., owned by Stephen Girard of Redding. It is 13 miles by road northwest of Redding, the last mile of which is not in good condition. According to Girard, the property was worked 40 years ago by Joe and Oliver Longfield, and the Minnesota Company bought it in 1891 or 1892. An arrastre, driven by water power from Spring Creek, was used in the earliest work, and later a 10-stamp mill was run for three months. Girard worked in the lowest adit level at the 900-ft. point years ago, and thinks that the total length is 1200 ft. At time of visit, he was cleaning it out and retimbering at the 500-ft. point. The level starts in a shearzone in a dike, probably alaskite, and follows this for 300 ft., at which point quartz starts to come in; and it widens to 4½ ft. at the 500-ft. point. Beyond this is a caved stope, which appears to have been 7 ft. wide. The ore is quartz carrying pyrite and chalcopyrite, with a strike of N. 75° to 80° W., and dip 70° N. The general formation is quartz diorite, intruded by dikes. On the unpatented claim, at a point 1300 ft. to the west of workings just described, a tunnel 350 ft. long, of which 250 ft. is open, follows an 18-inch quartz vein. The strike and dip are the same as those in the main working tunnel, and the vein may be the same one.

Bibl: State Mineralogist's Reports X, p. 635; XII, p. 255; XIII, p. 363; XIV, p. 794.

Molly Gibson (No. 73) is a group of seven claims in the Harrison Gulch District, Sec. 4, 5, T. 29 N., R. 10 W., held by John Kutz of Knob. According to the owner, there are nine veins on this group, opened by means of over 1000 ft. of tunnels. It was not visited during the present survey due to lack of time.

Mount Shasta Mine (No. 74) including one patented claim and two unpatented claims, total 52 acres, in Sec. 34, T. 32 N., R. 6 W., 10 miles west of Redding, is owned by H. H. Shuffleton of Redding. The Monitor and Lucky Boy of 20 acres each were surveyed for patent under M. S. 3582, but the patent was not completed. The deposit consists of two parallel veins, 50 ft. apart, of white quartz carrying pyrite and gold, in a dike of sheared alaskite porphyry in the quartz diorite. The main vein is 5 ft. wide and the parallel vein 4 ft. There is a recorded production of $178,000 from smelter shipments, and some additional production was made with a stamp mill that stood on the property; but there is no record of this.

Work on the seventh level, 465 ft. below the collar of the shaft, consisted of 140 ft. of drifting to the north on the west, or main vein, to a fault, and a little drifting was done on the east vein. Later H. O. Cummins sank the shaft 193 ft. deeper, cut a station, and started a crosscut, but quit late in 1913. After January, 1914, a watchman was kept on the property for several years. Surface improvements have since burned, and the mine is full of water. The owner has reports on the mine by H. O. Cummins and Guy M. Vail.

Bibl: State Mineralogist's Report XIV, p. 794. U. S. G. S. Bull. 540, pp. 39, 44, 46, 47–48.

Mountain Copper Co., Ltd., The (No. 75) Sec. 34, 35, T. 33 N., R. 6 W. Complete technical details of the cyanide treatment of gossan by this company in a 550-ton plant were given in the April, 1931, chapter of State Mineralogist's Report XXVII, pp. 129–38, when a good profit was being made from the operation. Since that time, the extraction has been improved, the grade of the ore has been raised, and additional ore reserves have been blocked out. Hence the operation is very successful. An additional leaching tank has been added to increase the daily capacity also.

Other operations of this company, production of pyrite at the rate of 500 tons per day, and the flotation of siliceous copper ores, have been described in State Mineralogist's Report XXII, pp. 154–160. During the time of high prices of copper, in 1929 and early 1930, the flotation plant was moved to the Iron Mountain No. 8 mine, and remodeled. Flotation concentrates were produced and were shipped to Tacoma for smelting. Copies of the chapters of reports cited are available at offices of this division at 25 cents per copy; hence no attempt has been made to describe details in the present report.

Bibl: State Mineralogist's Reports XIV, 769–70; XX, 440–45; XXII, pp. 154–60; XXVII, pp. 129–38. Bull. 50, pp. 70–78.

Mountain Queen (No. 76) is a group of six unpatented claims in Sec. 30, T. 32 N., R. 5 W., half a mile east of the old town of Shasta, which is six miles west of Redding by State highway. P. G. Parry of Shasta is the owner. He has recently sunk a 50-ft. shaft on a quartz vein 4 ft. to 8 ft. wide, and drifted both ways from the bottom for a total distance of 50 ft. In the faces of these drifts the vein is 4 ft. wide. The strike is S. 65° E., and the dip nearly vertical. The work was done with the idea of exposing the intersection with a second, but smaller,

FIG. 5.   Gossan quarry, Mountain Copper Company. (See description No. 75.)

vein which strikes S. 70° W., and has a nearly vertical dip. At a distance of 200 ft. west of the vein first mentioned, is a shaft on the smaller vein, 30 ft. deep. This second vein varies from a few inches in width at the surface to 2 ft. at the bottom. Parry has found his best ore here, but he says that a good shoot was found on the surface by former owners of the claims at the intersection of the two veins. He expects to cut this intersection with 20 or 30 ft. more of drifting from the 50-ft. shaft. The small vein is said to mill $30 per ton in free gold, and a sample cut across the larger vein is stated to have given an assay of $13.14 in gold. One hundred tons of ore are broken, ready for milling. A few tons were recently milled in a small stamp mill of 50-lb. stamps attached to springs, but this was found unsatisfactory; and the mill has been removed.

Surface cuts indicate that the veins extend for considerable distances beyond present workings, but other veins are probably present; and additional development work is needed to prove continuity. Country rock is quartz-diorite. One wall of one of the veins is a fine grained, dark grayish green dike rock. C. S. Laughton is leasing six acres on the northwestern end of this property, and some additional adjoining ground.

National Mine (No. 77) (Veteran Mine or Forbes Mine) is a group of seven unpatented claims in Sec. 23, T. 33 N., R. 5 W., 10 miles north of Redding, held by G. M. Sleezer of Redding. It is an old group originally located about 1869, and at one time a 10-stamp mill was operated; but this has been removed. Later shipments were made to smelters, and total production is said to have amounted to $200,000. A 1200-ft. adit level and 700-ft. winze from it have been inaccessible for 20 or 25 years.

Recently C. E. Crowell, a lessee, opened up some ore near the surface at a point which Sleezer thinks is farther north than any of the old workings. A shaft was put down 40 ft. on a vein averaging $2\frac{1}{2}$ ft. in width, then a total of 10 ft. of drifting was done on both sides of the shaft at the 25-ft. point, and the lower part of the shaft was filled. The ore is white quartz containing rather abundant pyrite.

At a distance of 200 ft. from the shaft is a 50-ft. drift adit on a 1-ft. quartz vein, from the outcrop of which Sleezer says he shipped 48 tons of $20- to $24-ore to smelters years ago. A similar vein is exposed in an assessment cut about 1000 ft. distant, near the portal of the old 1200-ft. adit. On another claim, the Veteran, there is some 250 ft. of work, 75 ft. of which is drifting on a 2-ft. quartz vein. There are some upper workings on this same vein also. At the portal of the lower tunnel are a compressor and 25-hp. gasoline engine, which have not been used recently.

Niagara Summit Mining Co. (No. 78), c/o W. D. Tillotson, Redding, owns a large acreage of patented mining claims and other patented land in Secs. 6, 7, 8, 17, 18, T. 33 N., R. 7 W., in the French Gulch District. L. W. Bunge of Los Angeles is president of the company.

A large amount of work has been done on these holdings, most of it on the Yosemite claim, and past production is estimated in the references given below at somewhere around $1,000,000. In an upper level called the O'Neil tunnel, which runs northwesterly through the center of the Yosemite claim, a small segment of the Niagara vein was found about 200 ft. southwest of the center of the claim. The Barnes tunnel starts at creek level, cuts through the northeast corner of the Yosemite, and runs southwest through the claim. A little ore was found in the southwestern part. The tunnel continues into the north part of the Comet claim. A recent proposal has called for running to the north at just about the center of the Yosemite claim at a point 2300 ft. from the portal of the Barnes tunnel, in a search for the Niagara vein at this lower level. J. H. Porter of French Gulch has maps of these workings. Equipment includes a 4 to 5-drill compressor of 75 hp., and a 10-stamp mill equipped with crusher, vanners, etc. Work has not yet been started on the project mentioned.

Late in 1931, James Blagrave and son of French Gulch held a lease on one claim of this group, and were mining surface material consisting of oxidized, decomposed diorite porphyry, which formed a layer of sub-soil, a few feet thick, and a few feet below the surface of the ground. The value was stated to be $5 per ton in gold. It was sledded to a 'Cannon-ball mill,' capacity of 6 tons in 24 hours, driven by an automobile engine. Treatment was by amalgamation.

Bibl: State Mineralogist's Reports X, pp. 636–37; XI, p. 50; XII, p. 253; XIII, p. 363; XIV, p. 793; XVIII, pp. 43, 296; XX, p. 15. U. S. G. S. Bull. 540, pp. 35, 37, 44, 67.

North Star (No. 79) and *Elkhorn* are two patented claims, 34 acres, in Sec. 15, T. 31 N., R. 6 W., adjoining the Potosi mine in the Muletown District. They are assessed to C. W. George of Redding. The Potosi vein, on which some good gold ore has recently been found, appears to run through the North Star. It is developed by means of two shafts, but these are not now accessible. The Spring Gulch Mining Co., which is now operating the Potosi, is reported to have recently optioned the North Star also.

North Star and *Virginia claims* (No. 80), 40 acres in Sec. 8, 9, T. 33 N., R. 7 W., held by Geo. W. Garwood of French Gulch, are optioned to E. L. Waight of San Francisco. A quartz vein averaging about a foot in width, with a steep dip, is found in a dioritic dike intrusive into Bragdon slate. Upper workings consist of a 30-ft. cross-cut to the vein and 300 ft. of additional workings of which 90 ft. is drifting on the vein. There is a small stope, 20 ft. high and 20 ft. long. The vein is entirely within the dike. Roughly 70 ft. lower in elevation is the present working tunnel, 320 ft. long, mostly in the slate. It is now (winter, 1932) being driven east on a 2-inch quartz stringer to develop the junction of this with the vein found on the upper level. At time of visit, two men were working with hand tools. Garwood states that the vein is the same as one found in the old Franklin mine adjoining.

Bibl: State Mineralogist's Report XVII, p. 523.

Old Diggings District (No. 81) see accompanying map of the district. See also: Reid, Central, Texas, Evening Star, Walker, Calumet. This old district, producer of several million dollars in gold-quartz ores, is located in Sec. 33, 34, T. 33 N., R. 5 W. and Sec. 3, 4, T. 32 N., R. 5 W., 10 miles north of Redding. Mills using amalgamation were operated here at one time, but later there was a demand for the siliceous ores as flux in the copper smelters; and this outlet proved more satisfactory. Hence the old mills were removed. A few years ago, the last of the smelters in the district shut down also, and these mines were left without an outlet for their ores, with the result that they have all since been idle.

The accompanying map indicates an interesting possibility in the way of grouping these old mines into a single large property with a central treatment plant, if sufficient new ore can be blocked out to justify its construction. Engineers who have investigated the records of some of these old mines have looked with disfavor on the declining value of the ores in the later operations. However, Logan [8] states that the quartz was worth $2 per ton as flux alone. Hence there was an incentive in the later operations to mine the lowest grades of quartz on a large-tonnage basis, rather than to do the development work necessary

[8] Logan, C. A., State Mineralogist's Report XXII, pp. 168 and 171, 1926.

to block out ore of a better grade. As a result of this, the records show that the last ore mined was low in grade. However, it seems probable that if the future development work is designed to locate pay-shoots, the grade of ore can be kept higher.

The district is very accessible, being just across the Sacramento River from the main line of the Southern Pacific Railroad, and only about half a mile distant. It is easily reached by means of 10 miles of road from Redding also.

Old Spanish (No. 82), formerly Deakin and Taylor, is a patented property of 160 acres in Sec. 31, T. 32 N., R. 5 W., four and a half miles west of Redding, held by Ralph P. Newcomb of Berkeley. A five-year lease with option has recently been taken by T. J. Burke of Redding, who states that quartz veins carrying good values in gold are found on the property with widths of one to four feet. Workings amounting to several hundred feet were made here years ago, and shipments of ore were sent to the Keswick smelter. These workings have all caved, and early in 1932, Burke had started to clean out some of them. Some free-milling ore was found at the surface, but the sulphides come in at a depth of only about 10 feet. The veins are associated with the intrusive contact of the quartz diorite into the meta-andesite.

Bibl: State Mineralogist's Reports XII, pp. 246–47; XIII, p. 353; XIV, p. 796.

Oro Fino (No. 83) is a property of two patented claims, M. S. 5343, and one unpatented claim in Sec. 34, 35, T. 32 N., R. 6 W., 10 miles west of Redding, via Shasta, held by Mrs. Edna Behrens Eaton of Redding. Dumps indicate that several hundred feet of work, perhaps a thousand feet or more, were done here. Parts of the dumps are quartz, apparently set aside as a second grade of ore. Caving on the surface indicates that stopes were driven from some of the drifts. Remnants of the veins that can be seen indicate a width of about a foot of mineralized quartz. To the north of these old workings is an outcrop of quartz 8 or 10 ft. wide, on which apparently the only work is a surface cut, 2 ft. deep. The outcrop can be traced for about 100 ft., but the width is not maintained for this distance. Workings are practically all closed by caving; and buildings have burned. Mrs. Eaton has reports on smelter-shipments to The Mountain Copper Co. on some 20 lots, which show gold values of $4.40 to $70.40 per ton. Silver averaged about one ounce per ton. Where the tonnage is given, it seems to have been about 10 tons per lot. Most of the shipments ran $20 and $30 per ton.

The geology is similar to that of the adjoining Mt. Shasta mine, which is described by Ferguson.[9] The country rock is quartz diorite with masses, perhaps dikes, of rhyolite. The latter is very fine in grain, and is flowbanded.

Bibl: State Mineralogist's Reports XI, p. 44; XIV, p. 796.

Oro Grande (No. 84) is a group of six unpatented claims, 100 acres, in Sec. 5, T. 31 N., R. 5 W., four miles west of Redding, recently held by Lewis B. Jennings, Graybar Building, New York City. With James A. Skene of Redding as superintendent, considerable equipment was installed here about 1930.

---

[9] Ferguson, H. G., U. S. Geol. Survey Bull. 540-A, pp. 41–43, 1913.

A network of quartz veins occurs here in the Copley meta-andesite. They all have steep dips, and the strikes are in various directions; but they do not appear to be close enough together so that the whole mass can be mined. Skene was selecting lenses or shoots in these veins, which he stated would run $4 to $7 per ton in gold. Widths vary from a gouge seam to 1 and 2 ft. of quartz, with occasional swells to 5 ft. Several thousand feet of open cuts, 2 ft. wide and 3 ft. deep, some old, some new, expose this network of veins in an area scattered over two claims—40 acres. According to Skene, before the owner mentioned above acquired the property, $15,000 was produced from a shoot 36 ft. long above the 50-ft. level. These workings were under water at time of visit. They were stated to consist of a 63-ft. shaft with 181 ft. of drifting to the south and 137 ft. to the north on the 50-ft. level, two crosscuts, each 20 ft. long, and a raise to the surface from the north drift. No work has been done on the property recently.

Equipment consisted of air-hoist, 2 cylinders, 6 by 8 inches, single drum, headframe, bucket, bin, car, No. 5 Cameron sinking pump, Ingersoll Rand single-stage compressor belted to 60-hp. motor and connection to lines of Pacific Gas and Electric Co. There was also a mill consisting of a 9 by 15-inch crusher, Straub rib-cone ball mill, 3 ft. by 3 ft., Burdan pan-amalgamator. Rated capacity was 15 tons per 24 hours through a 40-mesh screen.

Paymaster Mine (No. 85) is a group of six unpatented claims in Sec. 3?, T. 33 N., R 8 W., assessed to P. X. Johnson and others of Lewiston. Late in 1931, the Broken Hills Silver Corporation was reported to have an option on the property, but apparently this was not exercised. By road, it is 33 miles northwest of Redding, half State highway, and half mountain road. Several hundred feet of drifting on quartz veins, on each of several levels, have been done here, and the veins stoped out. At time of visit, the main level was a 100-ft. crosscut adit, from which 400 ft. of drifting had been done. A 28-ft. winze from this level was stated to be in very good gold ore. This showed 3 ft. of quartz, and iron staining extending out into the walls for a width of several feet more. This stained material, Copley meta-andesite, was stated to be ore also by R. G. Bennett, who held an option on the mine at the time. The strike is east and west and dip 71° north. A 200-ft length of the vein had been stoped 80 ft. between levels, plus 75 ft. to the surface.

Bennett had a crew of five men and was operating a two-stamp mill, 36-inch Straub ball mill, 35-ton Overstrom concentrating table, and crusher. Power was furnished by gasoline engines of 8 and 10 horsepower. Treatment was amalgamation. No figures on production resulting from this operation were available.

Philadelphia and Roosevelt (No. 86) are an unpatented claim and a fraction in Sec. 17, T. 33 N., R. 7 W., in the French Gulch District, 24 miles from Redding. They are held by Tony Alexson of French Gulch, and are optioned and leased to W. A. Vogt of French Gulch. The Washington, Halcyon, and Niagara adjoin.

W. A. Vogt, working alone, is removing high-grade lenses of oxidized ore from the diorite-porphyry. Both walls are of this formation, decomposed and soft, but flat slate underlies the deposit. The lenses are from a few inches to 4 ft in width, and produce 15 to 30 tons each of ore that is stated to run $100 per ton in gold. It is a soft ochreous ore, yellowish and reddish brown, some black probably due to manganese; and it pans well in free gold. Mining is done by picking only, except that occasionally a mass of slate will require blasting. Ore has been shipped to the two-stamp mill of Earnest Blagrave, mentioned under the heading 'Hobo' above. The only treatment is amalgamation, and some of the quartz contains sulphides rich in gold, which is not recovered. Thirty tons of ore were mined and stacked ready for shipment, at time of visit.

Phillips Group (No. 87), also known as Laurel, is in Sec. 14, T. 32 N., R. 9 W., on Spring Gulch, a tributary of Indian Creek, 45 miles west of Redding, of which two miles is trail. The claims are held by J. W. Phillips of Lewiston. A small quartz vein, which the owner states gives very good assays in gold, is developed by a series of adits. The strike is northwest and the dip is vertical. The highest adit, 53 ft. below the outcrop, or 222 ft. above the level of the creek, is 160 ft. long, and exposes the vein with an average width of 6 inches, maximum, 18 inches. The next adit is 57 ft. below, and has a length of 130 ft. It does not follow the vein, but has the same general direction. A third adit, 100 ft. still lower, is 250 ft. long, and exposes the vein with an average width of 9 inches. Part of this is oxidized ore, and part sulphide, pyrite and sphalerite. The lowest adit level is about 20 ft. above the creek, and at distances of 67 ft. and 122 ft. from the portal, two raises have been driven through to the level above. This work exposes the vein with widths of one and 1½ ft. There is some work on the opposite side of the creek also, part of which has caved. An arrastre was operated here years ago, but no ore has been extracted recently.

Bibl: State Mineralogist's Reports XIV, p. 894; XVII, p. 541.

Porcupine Mining Co. (No. 88), a partnership of J. L. Maulen, Mrs. Maulen, Geo. E. Fenby and B. F. Blazier, has recently held a group of unpatented claims in Sec. 35, T. 32 N., R. 6 W., and probably extending into adjoining sections in the next township south. The group is 10 miles west of Redding, and is near the Oro Fino. It included the Last Chance, Tom Cat and Condensed, acquired from Wm. Moore of Redding; and the Porcupine and other claims were located later.

Considerable equipment, buildings, water-pipe lines, grinding machinery and jigs, were installed with the idea that the entire formation in the vicinity, called 'porphyry' by the operators, but probably quartz diorite, contains enough gold to pay if worked on a large scale. Where these principal operations were contemplated, development work consisted of an adit, 20 ft. long, with a raise to the surface, and an open cut 15 ft. deep. On a detached claim, the Condensed, there was stated to be a 60-ft. incline and 70 ft. of drifting on a quartz vein. The latter was not visited. Only a small amount of material was milled; and the plant was then dismantled.

Potosi Land & Mining Co. (No. 89), c/o Bert W. Gibbs, Box 327, Oakland, holds 47 acres of patented mining claims in Sec. 22, T. 31 N., R. 6 W., in the Muletown District, 12 miles southwest of Redding. In 1931, Dr. H. C. Erno, W. R. S. Bewsher and Frank E. Gelhaus, all of Redding, took a lease and option on the property. About 1890, some production was made at the south end of the group from a 90-ft. shaft, which developed an ore-shoot, 100 ft. long and 4 to 16 inches wide of high grade ore, which was stoped out. Later work included the sinking of a 75-ft. shaft near the center of the group, and drifting 100 ft. north and 110 ft. south from the bottom. These workings are in good condition, but there were no ladders in the shaft at the time of the writer's visit.

The partners mentioned above installed a prospecting outfit consisting of Quincy portable type compressor driven by a 70-hp. gasoline engine, a small air hoist and a headframe for a new shaft, which was put down about 50 ft. Early in 1932, a small mill was constructed consisting of 3½-ft. Huntington mill, crusher, 6 by 8 inches, amalgamating plate, and small concentrating table of Wilfley type, all driven by a 10-hp. gasoline engine. This was run for a time on oxidized dump-ore, and ore from open cuts. Additional prospecting by Gelhaus then exposed a shoot of good ore on the surface just north of the 75-ft. shaft. He removed 16 tons of this and put it through the mill, with a resulting clean-up from the plates stated to be $1110. This strike brought about the sale of the lease and option to Spring Gulch Mining Co., which is reported to be installing a 50-ton mill using amalgamation and table-concentration. Dr. Chas. Harvey of Marysville and Redding is in charge.

The vein is exposed at intervals by surface cuts for a length of about 200 ft. with widths varying from 16 inches to 3 and 4 ft. Depth reached in the cuts is 10 ft. The ore is quartz containing pyrite where not oxidized; but in the cuts, it is practically all oxidized and the gold is free, and in very small particles. The bottom of the 75-ft. shaft is in the sulphide ore. The deposit is associated with the intrusive contact of quartz-diorite into Copley meta-andesite.

Bibl: State Mineralogist's Reports XII, p. 254; XIII, p. 363; XIV, p. 796.

Rainbow Prospect (No. 90) is a group of four claims in Sec. 32, T. 32 N., R. 8 W., 15 miles southeast of Douglas City by steep mountain road. It is held by Mrs. L. L. Kelley of Fresno. W. H. Collins has an option on the property, and has recently made some open cuts with the idea of milling the surface material. Older work is on a flat quartz vein, nearly parallel to the hillside, showing a few inches to 2 ft. of white, sugary quartz. About four years ago, W. B. Wilson put some of this through an arrastre with good results. The prospect was idle when visited.

Bibl: State Mineralogist's Report XXII, p. 24.

R. A. M. and Little Alice (No. 91) are two groups of four unpatented claims each, held by C. H. Foster and James McLaughlin of Redding, in Sec. 24?, T. 34 N., R. 9 W. They are on Montgomery Ridge, and a road forking from the Buckeye Ridge road near Minersville reaches them. At the time of visit, in June, 1932, C. H. Foster was opening and widening an old, caved tunnel on the Little Alice group. This was driven on a dike of soft, yellow, decomposed rock, which shows white spots, apparently originally feldspar crystals. In the dike are seams of quartz, one inch wide, which pan gold. Foster states that an average sample taken from the dump material removed in opening the tunnel assayed $12 per ton in gold. He was working at a point 40 ft. from the portal.

On the R. A. M. group, just to the west, a 40-ft. cut exposes a 3-ft. width of red and brown hematite, with some yellow limonite, containing bunches and small lenses of quartz that pan gold. Additional work is to be done here to locate the source of quartz piled on the dump from some cuts made years ago. Country rock is a dark-gray, fine-grained, basic appearing rock, where seen in the fresher exposure. It is all much decomposed, and fractures are heavily stained with reddish-brown oxides of iron. This country rock is common to both groups of claims.

The Little Alice group has recently been reported to have changed hands. Machinery including a Gibson mill is to be installed.

Rattlesnake Prospect (No. 92) is in Sec. 8, T. 31 N, R. 5 W., five miles west of Redding. There are several surface cuts on different

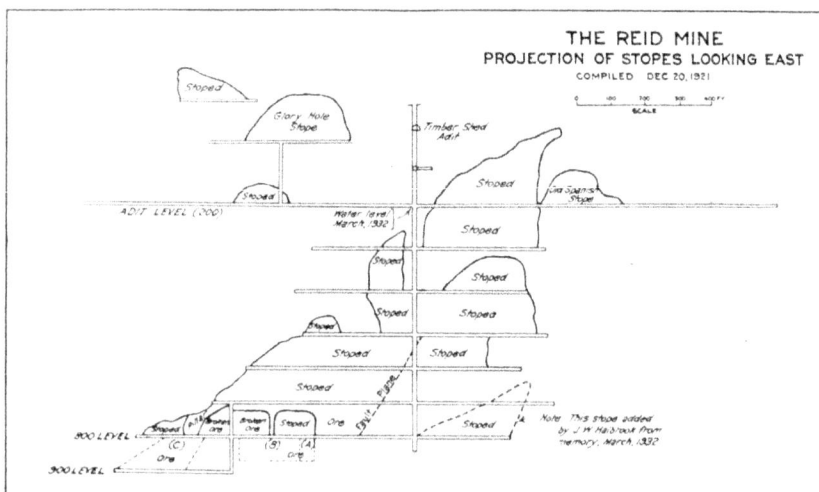

FIG. 6.   The Reid Mine, projection of stopes, looking east.   (See description No. 93.)

quartz veins in the Copley meta-andesite. Widths as great as 6 and 8 ft. were noted. Several hundred feet of crosscut adits have also been driven to prospect smaller veins. The owner, L. A. Davison of Delta, has recently been prospecting for pockets on the property. Equipment includes a small crusher, Gibson mill and gasoline engine.

Redding Consolidated of Nevada, see Boswell.

Reid Mine (No. 93), c/o D. V. Saeltzer, Redding, California, is located in Sec. 3, T. 32 N., R. 5 W., in the Old Diggings District (which see), a map of which is contained in this report. It is 10 miles north of Redding, and about one mile from the station, 'Central Mine,' on the Southern Pacific Railroad, with which it is connected by an aerial cable-tramway. There are 10 patented claims and fractions of a total area of about 140 acres.

The mine was operated continuously, from 1907 to the summer of 1919, by the owners, James and Harvey Sallee; and ore was shipped to the Mammoth smelter at Kennett, where the silica was particularly desirable as flux. When this smelter suspended operations in the summer of 1919, the mine was closed, as there was no longer a market for the ore. The mine was allowed to fill with water, and was idle and inaccessible until the fall of 1923, at which time it was optioned to Chas. H. Palmer, Jr., and associates of Los Angeles, who unwatered the workings, thoroughly sampled the ore exposed on the lower levels, and completed a small amount of development work on the 800 level, north of the shaft. The unwatering required about 90 days, and cost $7500 for labor, power, supplies and equipment. Harvey Sallee, who was then the owner, afterward operated the property for a short time, and shipped several thousand tons of siliceous ore to the Bully Hill smelter, and to the Kennett smelter during its last period of operation. After a few months, all operations were suspended, in the summer of 1925. The mine was again allowed to fill with water, and has since been idle.

The vein occurs in a fissure in andesite, and both walls are well defined, particularly the hanging wall, which is evidently a fault plane. The vein filling is white quartz of a sugary nature, and is easily broken and crushed, although it stands well while being mined. Values occur as gold with a small amount of silver, associated with pyrite and a small amount of chalcopyrite. These sulphides are usually disseminated throughout the vein, but occasionally occur in a more concentrated manner, in bunches and streaks in the vein, to produce high-grade ore. It is thought that gold tellurides occur also. The vein is offset in several places by cross-faults, with small displacements; but no faults of notable magnitude have been encountered. Most of the ore produced came from an oreshoot with a maximum length of 600 ft., having a steep rake to the north. The strike of the vein is north-south, and the dip 70° east. The width varies from a few inches to 25 ft., but the average width is nearer 10 feet. A large rhyolitic dike is found on the surface just to the east of the mine, but apparently it does not appear in the workings.

Equipment includes a head-frame, hoist-house, double drum electric hoist, three Ingersoll-Rand air compressors with a combined capacity of 1200 cubic feet of air per minute, machine and blacksmith shops, aerial cabletram and terminals, with ore-bins at the mine and at the railroad. Equipment is partially dismantled, especially the electrical equipment, and will require repairs before it can be used. The two-compartment, 1000-ft. shaft will also require some retimbering, particularly above the adit level.

Details of development are indicated on the accompanying 'Projection of Stopes' (Fig. 6), hence no attempt is made to describe them in this text. The blocks, A, B, and C, of partly developed ore between the eighth and ninth levels are taken from a report by H. W. Stotesbury, who was in charge of work done by the Palmer interests. Much of the other information herein contained on the Reid mine is derived from the same report. His sampling indicated that these blocks, if stoped to a width of 7 ft., contain possible ore to the amount of 14,290 tons of $4.04-ore, or if stoped to a width of 15 ft., 34,380 tons of $3.50-ore. Gold is the only metal considered in these estimates. This is a low grade of ore, but much of the ore from the mine has been twice and three times as high in grade, and even higher. For instance, from February, 1907, to February, 1909, 34,600 tons of a gross value of $9.64 per ton were shipped. In the later years of operation, when ore was being mined primarily for smelter flux, there was little incentive to do the development work necessary to locate high-grade shoots, as is discussed in greater detail in this report under 'Old Diggings District.' If development work is planned to locate such shoots, it seems probable that a good grade can be maintained, similar to that of some of the earlier operations. An interesting thing brought out in Stotesbury's report is that a parallel vein to that on which most of the work has been done is probably exposed on the lower levels, and ore-shoots might be developed in this second vein. The possibility of working the Reid in conjunction with other mines of the district is mentioned herein under 'Old Diggings District.'

Bibl: State Mineralogist's Reports XIV, p. 797; XVIII, p. 408; XIX, p. 11; XX, p. 15; XXII, pp. 176-78.

Sam Houston (No. 94) is a single unpatented claim in Sec. 34, T. 33 N., R. 5 W., in the Churntown Mining District, held by Roy Goll of Redding. It is 10 miles north of Redding, the last mile being trail at present; but an old road that reaches the claim by way of the Reid mine could be repaired at small cost. The deposit is a quartz vein associated with a rhyolitic dike in Copley meta-andesite. A surface cut shows 14 inches of quartz, from which Goll took a sample recently that is stated to have given an assay return of $10.10 in gold. Fifty feet lower is an 80-ft. tunnel crosscutting to an 18-inch vein, from which some ore is stated to have been stoped years ago and shipped to the Keswick smelter. The stope is caved. To the east, at a distance of 100 ft., is a 25-ft. adit on a steep vein varying in width from a few inches to 3 ft. At the portal is an 80-ft. shaft on the vein, showing that it has a very steep dip. Roughly 250 ft. lower in elevation than the bottom of the shaft is an adit, 380 ft. long, that was started to crosscut for the same vein, which has not yet been reached. Most of this adit was driven 25 years ago, but some of the assessment work has been done in it during recent years.

Bibl: State Mineralogist's Report XIV, p. 797.

Scottish Chief and Santa Clara, see Menzel.

Shasta Superior Mining Co., see Boswell.

Shasta View (No. 95) is a group of 21 unpatented claims and a mill-site in Sec. 26, 27, 34, 35, T. 32 N., R. 6 W., seven miles west of Redding, held for a number of years by F. A. Zimmerman of Redding. At the time of visit, it was stated that a corporation, Shasta View Gold Mining Co., Ltd., was being organized to work the property. D. M. Thies was stated to be president and Gilbert A. Nelson, secretary, address: 317 Clem Wilson Building, 5225 Wilshire Boulevard, Los Angeles. Homer I. Reynolds, Box 24, Redding, was on the ground as manager. He stated that a mining engineer's report made years ago shows a large body of commercial ore near the surface at a point 200 ft. ahead of the face of the main adit level. Workings on which this estimate was based have caved. Where the exposure was observed on the surface by the writer, a width of 6 or 7 ft. of white quartz was seen. This appeared to narrow farther ahead on the strike. The outcrop of another and somewhat larger lens of quartz was observed a few hundred feet beyond. The main adit level starts near creek level, and follows a shear-zone in a fine-grained dike, probably rhyolitic, in the quartz-diorite. Quartz lenses from a few inches to a foot wide occur in this shear-zone, and considerable disseminated pyrite is found throughout the shear-zone. The adit is 510 ft. long, 7 by 10 ft. outside timbers, and is timbered the entire length with drift sets of 12 by 12-inch square timbers. The direction is southwesterly. Equipment consisted of blacksmith shop, a number of other small buildings and a single-stage duplex compressor belted to an automobile engine.

When the property was visited in February, 1932, Reynolds was planning to drive the main level ahead, and install immediately a 120-ton mill using jigs, and to ship the concentrate to a smelter. At this time (Sept. 1932) the work has not yet been started; but another sale of the property has been reported. Workings on the opposite side of the mountain, facing the Mt. Shasta mine, were not visited.

Bibl: State Mineralogist's Reports XVIII, p. 256; XX, p. 15.

Snyder, see Clipper.

Spring Gulch Mining Co., see Potosi, also North Star.

Summit Mine (No. 96) (Summit Gulch Mine; Summit and Montezuma) is in Sec. 17, 18, 19, T. 33 N., R. 7 W., near the summit of the Tom Green road between French Gulch and Lewiston. The distance from Redding is 25 miles. There are eight claims in different groups, owned by various partnership combinations of J. H. Porter, L. W. Wheeler, M. A. Corliss, and Mrs. A. V. Fondahn, all of French Gulch.

Wheeler has been working on one claim of the group owned by himself. The formation is slate intruded by igneous dikes. A fissure vein of ribbon quartz with an average width of 2 ft., maximum 5 ft., strikes N. 25° E., and dips 60° E. He sank a winze on the vein from a 100-ft. drift adit, to a depth of 65 ft. This is stated to be in a pay-shoot, but it was full of water at time of visit. Another adit was later started at a point 170 ft. below the first, and driven for 400 ft. Several ore-shoots were found in this distance and were partly stoped out. The ore was milled in a 5-ft. Huntington mill using amalgamation, but difficulty was experienced in making a good recovery from the last ore treated, due to the sulphide content. Recently a 50-ft. winze has been sunk from this lower adit level, and 20 ft. of drifting done to the north. In these exposures, the vein varies in width from 6 inches to 3 ft., and is stated to run $6 to $8 per ton in gold. Additional drifting is to be done to pick up the better shoots.

On the adjoining Brown Bear claim, owned by J. H. Porter and the Wheeler estate, at least 2000 ft. of workings are open, exposing considerable ore in widths of 2 ft. to 4 ft., that is stated to assay $7 to $8 per ton. High-grade shoots have been removed by different lessees. An adit called the Galena tunnel, 540 ft. long, exposes some of this ore; and 150 ft. lower in elevation, the Barnes tunnel, 1012 ft. long, exposes more of it. From both workings, high-grade shoots have been stoped. From the Barnes level, a raise, about 150 ft. high, from the sides of which ore has been stoped at various elevations, is open. Below the Barnes level is a connection to the old workings of the Tom Green mine adjoining, adits of which have caved at the portals. A plan map of Wheeler's workings and sketch longitudinal projection of the older workings are on file in the Redding office of this division. J. H. Porter is one of the owners of the Tom Green group (which see).

Bibl: State Mineralogist's Report X, p. 641; XIII, p. 365; XIV, p. 799; XVIII, pp. 206, 256. U.S.G.S. Bull. 540-A, p. 61.

Sunny Hill Mining Co. (No. 97), c/o Mary Marsicano, 781 Green St., San Francisco, is the owner of a property in Sec. 1, T. 30 N., R. 8 W., seven miles west of Ono. The mine was a producer some 40 years ago from a rich surface shoot, from which shipments were made to smelters. Widths are stated to have varied from a very narrow seam to 4 ft. Later ore was milled on the property and cyanided. The cyanide equipment has been removed, but the 5-stamp mill is still there.

L. F. Barlow of Ono is driving a lower adit level, which it is stated will cut the vein at a point several hundred feet below the lowest of the old workings. It is in 800 ft., and 1200 ft. more will be required to reach the vein, according to estimates. The country rock here is a schist varying in appearance from dioritic material to serpentine.

Bibl: State Mineralogist's Reports XII, p. 257; XIII, p. 365; XIV, p. 799.

Sybil Mine (No. 98) (Shasta Hills, Accident, Anaconda) is a group of 20 unpatented claims in Sec. 7, T. 33 N., R., 7 W., of a total area of 360 acres, held by L. Von Krusze and G. A. Von Krusze of French Gulch. It is 26 miles from Redding by road, 18 miles highway, balance mountain road.

The deposit is typical of the French Gulch District—quartz veins associated with dikes of diorite-porphyry in slate. The quartz contains pyrite, sphalerite, galena, arsenopyrite, and free gold. The only level now open is called the sixth. Levels above this developed the vein at short intervals to the outcrop, 200 ft. above the sixth level. A shaft below the level, now full of water, is stated to be down 150 ft., and to have developed a shoot of ore 90 ft. long, average width 4 ft., with a variation in width from 2 ft. near the ends to a maximum of 9 ft. at the middle. The shoot has been stoped out. Where exposed on the bottom level, its assay value is stated to be $5.42 per ton. Ore extracted is said to have varied in value from $30 to $600 per ton. A 2-inch Cameron pump, running 8 hours per day will keep the water out of these workings. G. A. Von Krusze is driving a crosscut from the sixth level, by hand, in the slate, in search of a vein that he states was exposed on the No. 2 level. He has driven 75 ft. and expects to drive 50 or 60 ft. farther. Veins strike southwest and northeast, and are nearly vertical in dip. The best ore has been found where a slate contact comes in on the hanging wall of a vein. The intersection of such a slate-porphyry contact, near the shaft mentioned above, has a dip to the south, and gives the ore shoot a rake in that direction. Footwall is diorite porphyry.

L. Von Krusze has recently stoped, underhand, a small stringer near the shaft or winze, and treated the ore in a mill of the cannon-ball type, driven by a gasoline engine. Treatment was amalgamation. The stope goes down about 10 ft. to the water-level. A lower tunnel, called the Accident, has been started to tap the mine 540 ft. lower in elevation. It is 800 ft. long, and 400 ft. more will be required to reach the vein, according to the owners. The portal is caved. This tunnel is stated to expose a 4-ft. vein that will free-mill $8 per ton; and some stoping has been done on it.

The mine is connected to the lines of the Pacific Gas and Electric Co. Equipment includes a 12 by 12-inch Rix single-stage compressor, driven by 50-hp. motor, 15-hp. electric hoist, several small pumps, cars, track, blacksmith shop, bunk house, boarding house, and other buildings. The mill is equipped with a crusher and 7-hp. motor, 5-stamp battery, 1200-lb. stamps, with a 15-hp. motor, and a Senn vanner.

Bibl: State Mineralogist's Reports XIV, p. 777; XVIII, pp. 43, 138, 296, 408; XIX, p. 11; XXII, pp. 178–79. U. S. G. S. Bull. 540, pp. 68–69.

Texas Mine (No. 99) comprises 7 patented claims and a fraction in Sec. 33, T. 33 N., R. 5 W., in the Old Diggings District, 10 miles north of Redding. (See 'Old Diggings District' and map of that district). The property is owned by Mrs. Mary B. Garlick, 2243

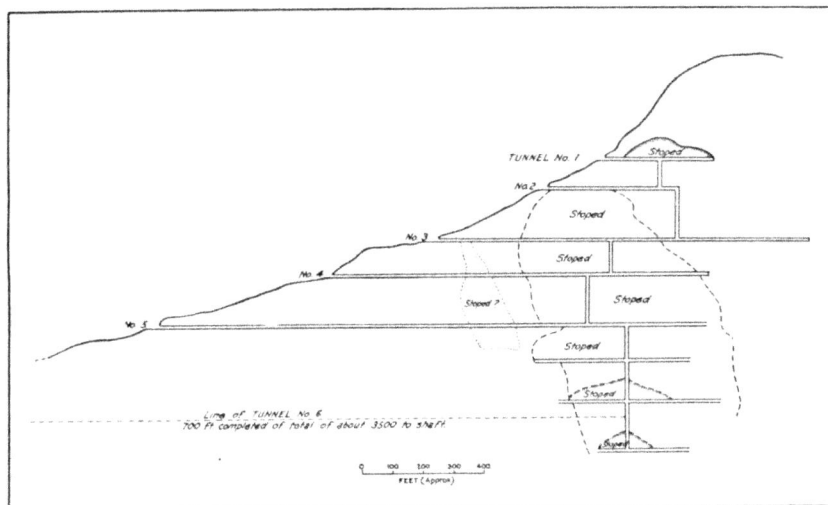

Fig. 7.   Sketch projection of underground workings of Middle vein, Texas Mine, from data furnished by W. D. Tillotson of Redding. Stopes are shown the same as on original, but Tillotson says that all within the dotted lines has been stoped. (See description No. 99.)

Franklin St., San Francisco, for whom W. D. Tillotson of Redding is local representative. It has been idle for a number of years, and portals of adits have caved, making workings inaccessible. Information given here is derived from State Mineralogist's Reports listed below and a report and notes furnished by W. D. Tillotson.

Gross production is roughly estimated at $750,000. According to notes of the late Dr. Garlick, who owned the mine, he was able to find receipts of sales of bullion, amalgam, concentrates, and ore totaling $403,747. However, he was unable to complete the record, receipts for various periods being missing.

The property is known to contain three different veins, but development has been confined largely to one of them, the Middle Vein, work on which is shown on the accompanying sketch-projection (Fig. 7). The width is reported as varying from 4 to 14 feet. A little ore was also produced from the East Vein near the surface. A 20-stamp mill of 825-lb. stamps, concentrators, canvas plant, and chlorination works was operated here for a number of years about 1890. Mining was done by hand, and 75 men were employed. Capacity was about 40 tons per day. The last work was done by Harvey Sallee, in 1922–23, and consisted of crosscutting in a search for segments of the vein beyond a fault.

The bottom of the mine appears to be in a lean zone, but additional development work might pick up other ore-shoots. The Reid mine, just to the south, has produced from much greater depths. The possibility of grouping these old mines for economical working has been mentioned under the heading, 'Old Diggings District'.

Bibl: State Mineralogist's Reports X, pp. 629–30; XI, pp. 43, 396–97; XII, p. 258; XIII, p. 365; XIV, p. 800; XVIII, p. 206; XIX, pp. 11, 58, 137; XXII, p. 179.

Tom Green Mine (No. 100) consists of four unpatented claims in Sec. 17, 18, T. 33 N., R. 7 W., adjoining the Summit Mine (which see) in the French Gulch District. The owners are J. H. Porter, Chas. Fox and Henry Carter, all of French Gulch. The mine has been a producer in the past, but portals of adits are caved at present. However, some of the workings can be entered through a connection with the adjoining Summit mine, the adits of which are on the opposite side of the mountain. Some of the Tom Green workings are shown on a sketch longitudinal projection of the Summit mine on file in the Redding office of this division. In the summer of 1932, T. W. Kenny and Geo. H. Barber were installing a 5-stamp battery of 200-lb. stamps to treat dump material. Power was to be furnished by a 7-hp. gasoline engine.

Bibl: State Mineralogist's Report XIV, p. 800.

Trinity Gold Mine (No. 101) (Prospect) comprises six unpatented claims, 120 acres, in Sec. 6, T. 31 N., R. 9 W., held by Harry Paige and S. G. Lovejoy of Douglas City and Dr. H. E. MacDonald of Redding. From Redding, it is reached by 40 miles of State highway plus 7 miles of dirt road to the mine, which is located on a steep slope draining into Browns Creek.

The deposit consists of quartz stringers and a silicified zone in schist, which is stated to contain gold enough to make mining on a large scale profitable, providing a sufficiently large tonnage can be developed. The ridge to the west of Browns Creek rises to an elevation of roughly 1000 ft. above the creek. At a point 650 ft. above the creek, a cut 70 ft. long and 3 ft. deep exposes quartz stringers and silicified schist for the entire length. One hundred feet below is a 65-ft. tunnel in the same material. A new adit, 50 or 60 ft. above the creek, was being driven at the 110-ft. point at the time of visit. All of the country rock is schist, some of it greenish and chloritic, some slaty. All of it is silicified and cut by quartz stringers; but these are much more numerous in the upper workings than in the part of the new adit inspected. It was planned to drive this adit at least to the 300-ft. point. An assay sheet of samples stated to have been taken by O. M. Wilson, shows 19 samples with an average value of $3.53 per ton in gold. These samples were taken at various points on the surface and in open cuts, scattered pretty well all over Trinity No. 2 claim, and in the upper tunnel, which is just east of that claim. Widths of samples are not stated.

Trinity Mohawk (No. 102) includes 6 claims formerly held by J. H. King of Redding, which are being purchased by Vest F. McClelland, J. L. Gow, and W. H. Wing, also 12 additional claims which have been located by the three last named and Ed. Johnson. There is also a 5-acre millsite. The location is in Sec. 26, 27, T. 34 N., R. 8 W., 34 miles from Redding, 18 miles of which is State highway, balance steep mountain road.

In a quartz fissure vein heavily stained with oxides of iron, the operators state that 10,000 to 12,000 tons of $10-ore with an average width of 5 ft. is in sight. The vein has a dip of 35° to the north. On the main adit level, there are 200 ft. of drifting on the vein and several hundred feet of crosscutting. An intermediate, 40 ft. above, has 150 ft. of drifting; and 30 ft. above that, is a level with 130 ft. of drifting and a small stope with a hole to the surface 30 to 40 ft. above. Snow blocked the road to the mine during the winter, 1931–32, and development work was carried on with supplies packed up the mountain from Lewiston. New drifting amounting to 125 ft., and raising amounting to 40 ft. was done. Country rock is slate, with possibly an intruded dioritic rock. There is also a fine-grained siliceous conglomerate on part of the holdings. Other veins are known on the property, but development has been largely confined to one.

·   At the time of visit, 30 tons of ore had been run through a recently constructed mill consisting of a 40-ton Forrester-Rexman rod mill with a 30-mesh screen. This was equipped with a 2 by 5-ft. amalgamating plate bent to fit within a few inches of the peripheral discharge of the rod mill. There were also inclined plates below, 4 by 6 ft. and 4 by 5 ft. An exceptionally high extraction by amalgamation alone is claimed but sulphides will, no doubt, be found as development proceeds. Power is supplied by a 32-hp. gasoline engine. Ten men were employed in mine and mill.

Water supply for the mill is brought from the opposite side of the mountain through a new 8500-ft. line of 2-inch casing. When this was being laid, a new vein was found on the opposite side of the mountain from the main workings, on the slope that drains into Little Papoose Creek. This had been opened in three places by means of shallow cuts, indicating a possible length between cuts of several hundred feet. The surface quartz panned well in free gold, fairly coarse. More extensive development work was to proceed immediately on this.

Bibl: State Mineralogist's Reports XVIII, p. 498; XXII, pp. 26, 32.

Truscott Mine (No. 103) (Martin Mine, including the Iron Mask) is a patented quarter section, NE ¼ Sec. 36, T. 33 N., R. 7 W., owned by Mrs. John Martin of Redding, and recently leased to J. C. Clemence and F. E. Brewster. It is 18 miles west of Redding by State highway and 2½ miles of dirt road.

Clemence and Brewster installed prospecting equipment consisting of an 85-cu. ft. portable Gardner-Denver compressor driven by a gasoline engine, and a 4-inch ventilating fan driven by a small engine.

They started at the face of an old 213-ft. adit, and drove about 300 ft. to a slate-porphyry contact, on which some old stoping can be seen on the surface. This porphyry is of the "bird's eye" variety, probably andesitic. They found the contact, but no ore, and the work was abandoned. An old mill building contains a 14-stamp Straub mill, rated at 20 to 30 tons per 24 hours, also a 5-stamp mill of 850-lb. stamps,

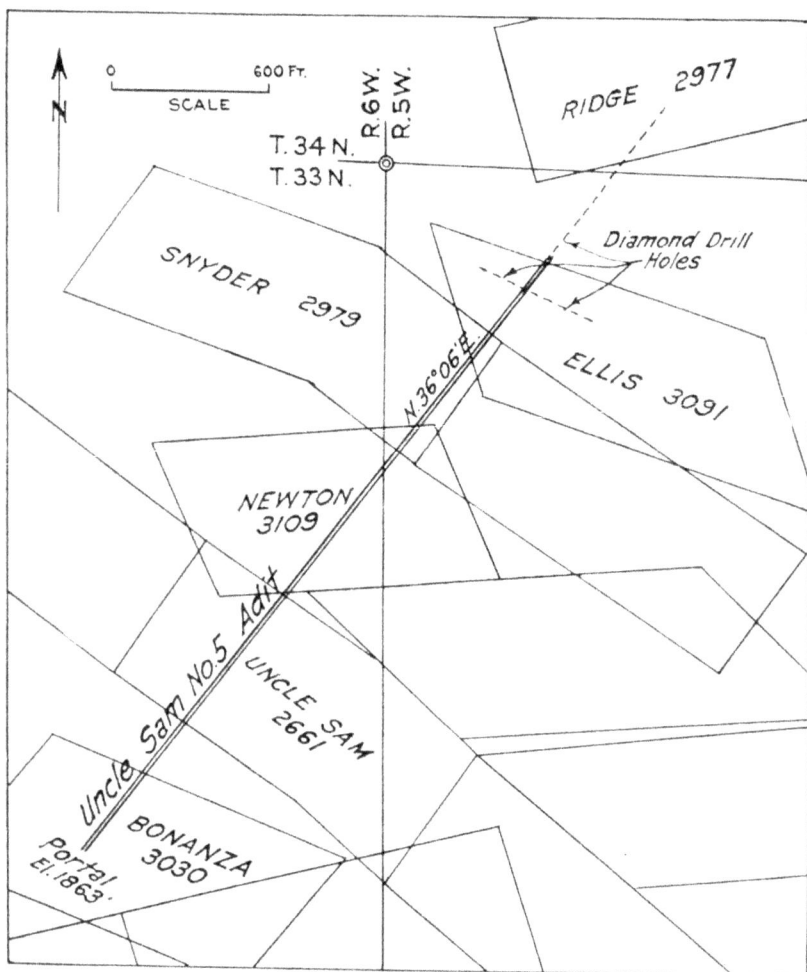

FIG. 8. Plan of Uncle Sam No. 5 adit and claims (with U. S. Mineral Survey numbers) through which it passes. (See description No. 104.)

amalgamating plates and a circular concentrator. The old steam engine that drove this outfit is in poor condition.

Bibl: State Mineralogist's Reports VIII, pp. 571-72; XII, p. 248; XIII, p. 356; XIV, p. 802. U. S. G. S. Bull. 540-A, pp. 48-49.

Uncle Sam Mine (No. 104), consisting of a large acreage of patented mining claims, in Sec. 1, T. 33 N., R. 6 W., Sec. 6, T. 33 N., R. 5 W., Sec. 36, T. 34 N., R. 6 W., is the property of the Dakin Co., c/o

C. C. Dakin, Redding. It is one of the well-known, old gold-quartz mines of Shasta County, having produced over $1,000,000 in several years prior to 1913, when it was operated by the Sierra Buttes Mining Co. Plan of workings and projection of stopes are to be found in State Mineralogist's Report XIV, p. 801.

About 1923, the American Zinc, Lead and Smelting Co. of St. Louis drove the No. 5 adit to a point just beyond the north side-line of the Ellis claim (which see) and did several hundred feet of diamond drilling, in different directions, from points near the face. This work was done in a search for extensions or segments of the great bodies of copper ore known in the adjoining Mammoth mine of the U. S. Smelting, Refining and Mining Exploration Co.

In driving this adit, several quartz veins of good size were cut in holdings to the north of those of the Uncle Sam; but they were not developed. These veins are attracting attention at present as good gold prospects. The difference in elevation between the adit level and the surface affords backs of 1000 ft. and more.

The adit has recently been opened to such an extent that the face can be reached for inspection, but several small caves must be cleaned up to make it serviceable. The 6-mile road up the mountain from the

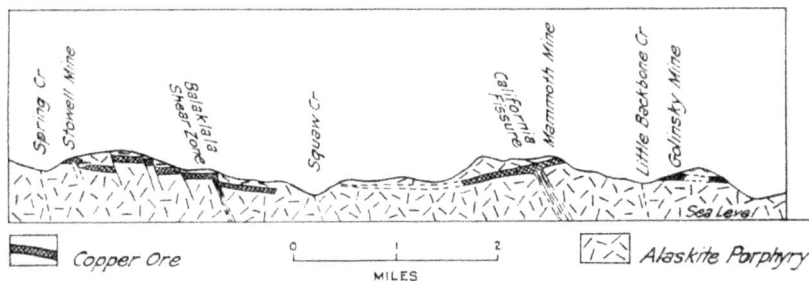

FIG. 9.   Idealized longitudinal section, looking northwest, of a part of the Shasta County Copper Belt, adapted from an unpublished section by M. E. Dittmar. Deposits of gossan are found at the surface, where ore-bodies outcrop. (See description No. 5 and No. 105.)

Squaw Creek bridge near Kennett is entirely washed out, and many stretches must be entirely rebuilt. A 10-stamp mill in a partially wrecked condition stands near the portal of the No. 5 adit.

Bibl: State Mineralogist's Reports X, pp. 310, 639; XI, pp. 47, 395, 398; XII, pp. 258–59; XIII, pp. 367–68; XIV, pp. 802–03; XIX, p. 56; XX, p. 15.

United States Smelting, Refining & Mining Exploration Co. (No. 105) (Mammoth Copper Mining Co.), 1 State Street, Boston, Mass., owns thousands of acres of mining claims in Sec. 2, 3, 29, 31, 32, 33, T. 33 and 34 N., R. 5 W., near Kennett. A copper smelter of 2200 tons daily capacity was formerly operated at Kennett, but it has recently been junked. A description of the copper mines and map of the holdings are contained in the April, 1926, chapter of State Mineralogist's Report XXII, copies of which are available at offices of this division at 25¢ each.

Interest has recently been shown in regard to the possibility of producing gold from the gossan here by the process used by the Mountain Copper Co. Apparently the Mammoth gossan deposit is much smaller than that of the Mountain Copper Co., but the necessary development work to actually determine the tonnage is lacking. An engineering examination of the gossan with rather extensive sampling was made about 1931, but the company will not give out the results of this.

The company owns also two patented claims in the Old Diggings district (which see), from which some gold-quartz ore has been produced from shallow workings. These claims are of interest as possible members of a large group that could be formed at Old Diggings for economical mining and milling of the ore.

Bibl: State Mineralogist's Reports XIV, pp. 767–69; XX, pp. 435–38; XXII, pp. 150–54. Bull. 50, pp. 95, 97.

Utah and California Mine (No. 106) (Walker Mine), consisting of 12 patented claims and fractions in Sec. 3, 4, 10, T. 32 N., R. 5 W., in the southern part of Old Diggings District (which see), owned by Walker Bros. of Salt Lake City. It was formerly a producer equipped with a 10-stamp mill, but has been idle for years; and there is no equipment. At this time (Oct. 1932) the lowest adit level is being opened for an engineering examination and possibly new development work.

Bibl: State Mineralogist's Reports X, pp. 630–31; XI, p. 397; XII, p. 259; XIII, p. 368; XIV, p. 803.

Venecia Mine (No. 107) (Eastman Consolidated Mines Corporation) comprises 13 claims, of which one is patented, in Sec. 3, T. 33 N., R. 8. W. The distance from Redding is 34 miles, half highway and half mountain road crossing a 4000-ft. summit. Paulsen Bros. of Lewiston are the owners, and have given a lease and option to the corporation named above, which had two men at work at time of visit. Three quartz fissure veins with vertical dip, varying in width from a few inches to 5 ft., occur. The ore is base, containing quartz, gouge and iron, zinc and lead sulphides. Country rock is Copley meta-andesite and quartz diorite, and a little slate.

Development work consists of a No. 3 level with 200 ft. of workings open, largely drifting on the Venecia vein, balance caved; also a No. 5 level with 1000 ft. of workings of which 500 ft. is drifting on a small fault vein showing gouge and a few inches of quartz. A crosscut was being driven on the No. 5 level to the Venecia vein at time of visit. An intermediate between these levels is 350 ft. long. W. G. Carlson, who was in charge of the work, estimated 5000 tons of ore of a good grade in sight.

Equipment consists of an 8 by 8-inch Sullivan compressor with a 60-hp. engine, machine drills, cars, rails, pipe, etc.; also a 6-ft. Huntington mill. None of the machinery was in service at time of visit.

Bibl: State Mineralogist's Reports XVIII, pp. 207, 497; XIV. p. 900.

Veteran, see National.

Victor, see Harrison Gulch.

Walker, see Utah and California.

Washington Gold Mining Company (No. 108) is controlled by Dr. Geo. Grotefend of Redding. This patented property is in Sec. 16, 17, T. 33 N., R. 7 W., and covers most of the hill between the forks of French Gulch. It was located in 1852, and is one of the oldest quartz mines in the State. For 50 years it was operated on and off, and produced a total amount of gold variously estimated at between $1,000,000 and $2,000,000.

Through a series of six adits, it has been developed at various elevations from near creek level, elevation 2000 ft., to a point near the top of the ridge, elevation 2900 ft. The upper workings are partly in the Bragdon slates with the dikes of intruded soda granite-porphyry characteristic of the French Gulch district. Copley meta-andesite is found in some of the lower workings. Two main veins have been worked, one striking north and dipping east, the other striking a little north of east and dipping north. Small stringers found in the vicinity of dikes have also yielded considerable ore. Stoping widths of 5 ft. and more have been found on the north-south vein, which was the largest. Ore is principally white quartz with small quantities of galena, sphalerite, pyrite and arsenopyrite.

Recently (summer, 1932) the I (eye) level, 270 ft. above the creek, has been cleaned out and retimbered for a distance of 600 ft. Some 50 ft. beyond this is a small cave under a raise, and 50 ft. beyond the level is closed by caving. Some small quartz veins are seen in branches of this level. On the H level, 220 ft. above the I, 1000 ft. or more of work is open, and quartz veins with widths of 2 to 5 ft. are seen. The G level is caved at the portal. The F or Trimbath level, 715 ft. above creek level, is partly open, and the north-south vein, which was the principal producer in the early days of the mine, can be seen. According to Wm. Blagrave of French Gulch, a former lessee, the ore-shoot on this vein had a rake to the north, and development work to the north on some of the lower levels should pick it up. Some hundreds of feet of intermediate levels are also open. A detailed mapping of all open workings, with plotting strikes and dips of all veins, faults, dikes and contacts should be of great assistance in locating new ore in this old mine.

A cut on the surface, high on the ridge, recently yielded 12 tons of ore, from which Dr. Grotefend states $725 in free gold was extracted. There is an old five-stamp mill on the property.

Bibl: State Mineralogist's reports X, pp. 635-36; XI, p. 50; XII, p. 260; XIII, p. 368; XIV, p. 804; XVIII, p. 43; XIX, p. 11; XXII, p. 179–80. U.S.G.S. Bull. 540, pp. 35, 37, 44, 64–66.

Woodfill, see Beaver.

Yankee Jack Mine (No. 109) (Yankee John) is a group of three unpatented claims in Sec. 17, T. 31 N., R. 5 W., 6 miles southwest of Redding by road. It is held by John F. Reese and W. L. Hill, both of Redding.

The 300-ft. shaft is kept unwatered to the 200-ft. level by the owners. This requires bailing for 20 minutes per day. The 300-ft. level has not been accessible recently. According to maps and reports, all work on the 300 was to the northeast; while on upper levels, ore recently discovered was to the southwest of the shaft. On the 200-ft. level, on the north side of the shaft, there is a 35-ft. crosscut to the main vein, on which is a 100-ft drift to the west. The average width is 5 ft., with a maximum of 8 or 9 ft. where the vein splits near the west face. Owners state that this ore runs $10 per ton in free gold, or $12 to $13 total including sulphides. The drift also extends to the east of the crosscut, exposing smaller widths, but the grade of part of this is stated to be higher. In the crosscut, at a point 15 ft. north of the shaft is a drift on a parallel vein 2 ft. wide, which the old Yankee John company stoped out, about 1915. This vein is said to have been considerably wider on the 160 ft. level, which was not open at time of visit. A third working on the 200 level is a crosscut in an easterly direction from the shaft. At 35 ft. from the shaft, this struck a stringer from which a pocket was taken by means of a 20-ft. winze. At a distance of 180 ft. from the shaft, a vein is exposed which is stated to assay $7 to $12 per ton for a width of 10 ft. Ore in this vein is a soft, partially oxidized quartz in contrast to that of the vein where most of the work has been done, which is a hard white quartz containing pyrite, and minor quantities of other sulphides. The soft ore has been exposed for only a few feet along the strike, but it appears to be a promising prospect.

On the 100-ft. level, the old crosscut was driven 16 ft. northwest from the shaft, then turned to the west for 100 ft., not on the vein. Reese and Hill have crosscut to the north 25 ft. from the old face, and have struck the main vein, then drifted west for 45 ft. They extracted 50 tons from this drift that is stated to have yielded $2,000 in free gold. There is stated to be considerable gold in the sulphides which went into the tailing also. Ore above the 70-ft. level was stoped out years ago. Country rock is Copley meta-andesite, but slate is to be seen on the surface just to the east of the shaft. Andesitic dikes with larger feldspar phenocrysts than those of the Copley meta-andesite are found on the north side of the main vein, but appear to strike at an angle with the vein.

Equipment consists of a 9 by 12-inch single stage compressor, and a hoist, both driven by one 15-hp. gasoline engine. The 10-stamp mill that formerly stood on the property has been removed. The ore treated recently was hauled to a mill on another property of Hill's, 10 miles north of Redding.

Bibl: State Mineralogist's Reports XVII, p. 524; XVIII, p. 495; XIX, p. 11.

GOLD (Placer)

Bigelow Ranch Placer (No. 110) consists of 100 acres of unpatented placer claims in Sec. 31, T. 32 N., R. 9 W., on Browns Creek, owned by Pearl Bigelow and Harvey Bigelow of Douglas City. In 1931, this was optioned to Harry Holloway, who sank several prospect shafts in the gravel, and washed out a few prospect pits with water pumped from the creek by means of gasoline engines. Recently (1932) the property has been idle.

Big Bend Placer Claim (No. 111) in Sec. 17, T. 33 N., R. 8 W., one mile north of Lewiston, and other property in Sec. 19 and 20, on the other side of Trinity River nearer Lewiston, have recently (May,

1932) been optioned by the owner, J. W. Phillips of Lewiston, to G. H. Harrison and brother of Watsonville. The total area of property controlled by Harrison Bros. amounted to 305 acres. An experiment was conducted here in pumping water from Trinity River to supply an ordinary giant for hydraulicking the gravel. The installation consisted of a Byron-Jackson double-suction centrifugal pump, built to order, and capable of delivering 4600 gallons per minute through 800

FIG. 10. Pump for hydraulicking at Big Bend placer. (See description No. 111.)

ft. of 15-inch pipe and a 3½-inch nozzle, with a pressure equivalent to a 320-ft. head. Power was furnished by a 300-hp. Westinghouse motor, 1760 r.p.m., connected to lines of the Pacific Gas and Electric Co. From a mechanical standpoint, the installation appeared to be a success; and a good stream under high pressure was delivered at the giant. However, the property was idle when visited in the fall of 1932. Some of the obvious difficulties were the lack of sufficient water to wash the gravel in the sluice, and the presence of a stratum of cemented gravel several feet thick next to the bedrock. At one time, the operators were considering the installation of a second pump to deliver water at low pressure for washing. The main consideration in an installation of this kind is whether or not the gravel contains enough gold to pay the operating expenses plus a proper return on the investment. Gold content of the gravel should be determined in advance of the installation of equipment by carefully prospecting the deposit through numerous holes drilled or sunk in a systematic way clear through the gravel to the bedrock. Operating expenses on such an installation are bound to be very high on account of the excessive power consumption due to the relative inefficiency of a stream of water in moving gravel as compared with a mechanical device such as a bucket-line on a dredge.

Bibl: (Phillips) State Mineralogist's Report XIII, p. 459.

Blue Gravel Mine (No. 112) is on land purchased by the City of Redding for the airport and lies just to the southwest of the latter.

About 1915, some good placer gold was found here by drifting methods. Total production is estimated at roughly $20,000. Of this, $7,000 was produced under the supervision of W. L. Hill of Redding. Some production has recently been made by George Dix, a lessee. Part of the work was done by hand, and later a ½-yard gasoline shovel was tried.

Buckeye, see Majestic.

Clear Creek Placer Company (No. 113), a Washington corporation, is purchasing from James E. Paige 80 acres of placer ground on Clear Creek in Sec. 34, T. 32 N., R. 6 W. Directors of the company include C. C. Merriam of Seattle and S. C. Brady of Yakima. Under the supervision of G. L. Covington an attempt was made to mine the gravel with 1¼-yard Thew gasoline shovel, and to wash the gravel in a machine mounted on skids, and designed to be pulled ahead by the shovel as it advanced. This machine had been altered several times but was still in the experimental stage when last seen by the writer. Gravel was classified by means of grizzlies to sizes above and below 1½ inches. Oversize was stacked by means of a belt-stacker, and under size was treated on screens and carpet-tables to recover the gold. No adequate provision had yet been made for disposal of tailing at the lower end of the machine, which was interfering with its operation. The addition of a bucket elevator or sand and gravel pump was planned to overcome this difficulty.

Diving Bell Mine (No. 114) is a property of 40 acres in Sec. 26, T. 32 N., R. 5 W., just northwest of Redding, on the Sacramento River.

FIG. 11.   Diving Bell Mine, near Redding.   (See description No. 114.)

In 1932, L. Wells and Wm. A. Wells attempted to mine from the bottom of the river. Gravel of a depth of 12 feet was removed with a ½-cubic-yard clam-shell excavator set up on the north bank of the river (see photograph). This gravel was sluiced with water pumped from the river and yielded a little gold. The principal yield was expected to come from cleaning the hard, irregular bedrock by a man working in a diving suit. The excavator was principally for the purpose of removing enough of the overburden so that an area of the bedrock would be clean enough for the diver to work. Apparently excavating machinery of a greater range than that used is needed to accomplish this. James E. Carney of Redding is the owner.

Douglas City Group (No. 115) consists of eight claims, 820 acres, held by Dr. D. B. Fields of Weaverville in Sec. 1, 12, T. 32 N., R. 10 W. and Sec. 6, 7, T. 32 N., R. 9 W., covering gravels along the south bank of Trinity River between Indian Creek and Redding Creek.

According to a report of R. C. Eisenhauer, furnished the writer by Dr. Fields, 10 or 15 acres have been mined off at the west end of the property with results stated to be good. Work by ground sluicing and rockers was followed by small-scale hydraulic operations with water under low pressure. Some development work has also been done by B. R. Brown of Weaverville. Tunnel No. 1, 40 ft. long, on the Old Crow claim, is stated to have shown gravel of a value of 11¢ per cubic yard. Tunnel No. 2, 110 ft. long, starting at the north rim-rock on the Barley Field claim, is stated to have exposed gravel of a value of 11¢ to 14¢ per cubic yard. At 100 ft. from rim-rock, a winze was sunk 40 ft. without striking large boulders or bedrock. On the Silvia claim, near the top of the ridge, a 40 ft. shaft was sunk with reported values of 11¢ per cubic yard.

The late Pierre Bouery, who had charge of operations at the La Grange mine for 15 years, was working on a report on the Douglas City group shortly before his death. According to a partial report by him, a copy of which was furnished the writer by Dr. Fields, he thought that there is gravel enough here to make a very large hydraulic mine, providing all parts will yield the values indicated by development work mentioned. However a systematic campaign of development work will be required to definitely determine the yardage and the average gold content per cubic yard.

Water to the amount of 3000 miner's inches could be put on the property for a part of the year by extending the ditch from the Indian Creek and Panwocket (which see), six miles to the southeast. A head of 500 to 600 ft. would be available. For large-scale operations additional water will probably be needed.

Gardella Dredge on Clear Creek has been dismantled.

Gas Point Dredge (No. 116) (formerly Ogden and Wilson) is in Sec. 3, T. 29 N., R. 6 W., on the Roaring River just above where it flows into Cottonwood Creek, near Gas Point, or 15 miles by road west of Cottonwood. Chas. F. Staheli and J. L. Cerney of Anderson have operated the dredge on and off recently, and some production of both gold and platinum has resulted. The dredge is 75 ft. long by 36 ft. wide by 6 ft. deep and draws about 3 ft. of water. Power is furnished by an oil engine of 60 hp. burning six gallons of distillate per hour. There are 32 buckets of 2½ cubic foot capacity. Water for washing is furnished by an 8-inch Yuba pump. Oversize from a trommel with ⅜-inch holes is stacked by a belt conveyor, and undersize goes to a sluice with riffles. Capacity is 400 to 500 yards per 10-hour shift. The gravel that has been dredged is 8 to 12 ft. deep, and lies on a bedrock of volcanic material only a foot or two thick. Below this there is more gravel which has not been dredged, nor has it been thoroughly tested for values.

Gold Bar Dredging Corporation (No. 117). (A *Gardella* dredge burned at this locality some years ago.) This is a new company which has leases on 1000 acres of land on the Trinity River from the bridge at Lewiston down stream for five miles, in Sec. 18, 19, T. 33 N., R. 8 W. and Sec. 13, 23, 24, 26, T. 33 N. R. 9 W. The distance from Redding, the nearest railroad point, is 40 miles. Floyd R. Marsh is president, Chester E. McCarty is secretary, and Herbert Olson is superintendent. The home office is at 725 Yeon Building, Portland, Oregon.

A new dredge has been constructed, mostly under the supervision of E. E. Chenoweth, who was formerly in charge. The main hull, less overhang is 79 by 44 by 7 ft. The dredge is of the stacker type, and carries a bucket-line of 45 buckets of 7 cu. ft. capacity each, and digs to a maximum depth of 31 ft. below the water line. Gravel is dumped into a hopper, where it is washed with water under high pressure from two nozzles, then goes to a 24-ft. revolving screen, 4 ft. in diameter, where it is again washed with water under high pressure from several nozzles pointing in different directions. Largest boulders can be pushed off from the stacker-belt at the bottom, where the revolving screen discharges. The balance of the oversize is stacked. Undersize from the screens, which are drilled with holes from $\frac{1}{4}$ to $\frac{5}{8}$ inches in diameter, is treated on 3000 square feet of gold tables. These are arranged in the form of inclined sluices, 32 inches wide, equipped with riffles of 1 by 1-inch angle iron with a flat side turned to the top. The dredge is held in digging position by headlines of $1\frac{1}{2}$-inch steel cable only. No spuds are used. Electrical equipment is all first class, wired in conduit and with metal terminal boxes. It includes an automatic contactor system for the controller to pick up the load gradually. Power is delivered by the Pacific Gas and Electric Co. at 4400 volts, and is stepped down on the boat to 440 volts with three 100-kw. transformers. Motors are 150-hp. main digging motor, 30-hp. on 8-inch pump, 50-hp. on 10-inch pressure pump for nozzles in screen and hopper, 40-hp. for 7 drum winch and hoist. The dredge is equipped with two fire-fighting systems, water hydrants connected to pressure pump, and large soda-acid extinguishers, one of which is mounted on a cart. The cost of the dredge, not counting some metal parts of the old dredge, which were utilized, was $82,000.

The initial cleanup of a run started on August 16, 1932, of seven days, counting actual running time only, is stated to have returned 18¢ per cubic yard from 18,000 yards. This came from a bar that was dredged to a width of 300 ft. and a depth of 15 to 20 ft. The dredge is run three shifts per day with a total crew of 22 men. Drilling of the holdings of the company is stated to have shown an average gold content of the gravel of 17¢ per cubic yard.

Bibl: (Gardella Dredge) State Mineralogist's Report XVIII, pp. 297, 496, 601, 734.

Hook and Ladder (No. 118) is an old hydraulic mine in Sec. 5, 6, T. 33 N., R. 9 W., just north of Weaverville, which has been worked intermittently in a small way for many years. Water was bought from the La Grange Company for a short run of two months in 1929. In the summer of 1932, Trinity Exploration Co. attempted to mine the gravel here with a power-shovel, and treat it in a machine consisting of a revolving screen, sluice and belt-stacker (see photograph). The property was idle when visited.

Bibl: State Mineralogist's Reports XIV, p. 907; XXII, pp. 37, 54.

Igo Placer Mining Company (No. 119) has been prospecting 30 acres of gravel acquired from F. H. Russell, formerly part of his Lost Channel mine, in Sec. 34 ?, T. 31 N., R. 6 W., near Igo. John Ringus, 1276 Folsom St., San Francisco, was stated to be the owner, and C. O. Heynen, 400 Tesla St., Sausalito, Cal. was in charge of the prospecting. This consisted of advancing one tunnel 20 ft. and another 30 ft., and was stated to have exposed gravel of excellent grade. Several hundred feet of workings are open in the gravel in an old tunnel called the Stiller tunnel. Bedrock is a biotite-diorite, with a layer adjacent to the gravel usually decomposed.

FIG. 12. Washing machinery at Hook and Ladder Mine. (See description No. 118.)

Indian Creek and Panwauket Group (No. 120) comprises 10 claims, 913 acres in Sec. 23, 25, 26, 27, T. 32 N., R. 9 W., held by Dr. D. B. Fields of Weaverville. It is reached by seven miles of dirt road turning from the State highway at Douglas City. Equipment includes ditch of 3000 miner's inches capacity with water available during winter months, and as late as July in years of fairly abundant precipitation; also pipe, giants, and a number of buildings. Some extension of the ditch, say two miles, will be needed to reach the larger areas of gravel. During recent years Wm. Gribble and Milford Gribble have leased the property and have worked it on a small scale.

According to the late Pierre Bouery, who operated the La Grange hydraulic mine for 15 years, when it was running on a very large scale, the water from this property should be taken to the Douglas City group (which see), which is the more important of the two.

At Indian Creek, stream gravels in terraces above the creek have been mined for depths as great as 30 feet. They rest on a bedrock of sedimentary origin, sandstones and some conglomerate beds with an appearance very similar to that of the stream gravels, but cemented to some extent. No doubt, much of the stream gravel is derived from these beds of older conglomerate. These probably belong to the Horsetown formation (Cretaceous), but some Tertiary beds may be present also.

Bibl: State Mineralogist's Reports XIV, p. 907; XXII, p. 38.

FIG. 13.   Indian Creek placer.   (See description No. 120.)

FIG. 14.   La Grange Mine.   Small pit on upper rim, worked in 1929.   See description No. 121.)

La Grange Placer Mines, Ltd., (No. 121), a Delaware corporation, is a new company that has recently acquired the large property of the La Grange mine west of Weaverville. Thousands of acres are held in Sec. 3, 4, 9, T. 33 N., R. 10 W., and on Musser Hill to the northeast;

also water rights on Stuart Fork of Trinity River and various other streams between there and the mine. The large-scale operations at this mine have been described in State Mineralogist's Report XXII, chapter for January, 1926, copies of which are available at offices of this division, and in the other references listed below. The 11 miles of flume that brought water from Stuart Fork must be rebuilt and the additional 18 miles of ditch must be extensively repaired before the mine can be operated on the former scale. Much new equipment will be required also. A small amount of hydraulicking on the high rim of the old pit has recently been done with water brought by pipe-line from West Weaver Creek, by Milan Senger of Weaverville. Dr. Lyman Stookey of Los Angeles is president of the company, and Byron Stookey is secretary.

Bibl: State Mineralogist's Reports VIII, p. 638; X, p. 702; XII, p. 311; XIII, pp. 452–53; XIV, p. 908; XVIII, p. 257; XX, p. 182; XXII, pp. 39–42; Bull. 92, p. 94.

FIG. 15. Lewiston Dredge, partially dismantled. Bucket-line and stacker on ground behind. (See description No. 122.)

Lewiston Dredge (No. 122) (Placer Development, Ltd.; Lewiston Dredging Co.; Metals Exploration Co.) was shut down and partially dismantled in 1932, the stacker and bucket-line being removed and taken ashore. Its operations have been described by Lawrence K. Requa in a publication of the U. S. Bureau of Mines.[10] The dredge operated on a property of 500 acres of patented land, extending for about three miles along the main Trinity River and Stuarts Fork. The location is nine miles north of Lewiston or 50 miles from Redding, the nearest railroad point. To the end of 1931, approximately 10,000,000 cubic yards of gravel had been dredged.

The property was sampled in three stages by different owners, 100, 60, and 27 churn-drill holes being put down respectively, and a few

[10] Requa, L. K., Description of the Property and Operations at the Lewiston Dredge, Lewiston, Calif., U. S. Bureau of Mines Information Circular 6660, November, 1932.

shafts in addition. Drilling was done with a steam-driven portable rig (Keystone), and the last drilling, done in 1928, cost $7.23 per foot, including the salary of the engineer in charge. The cost of a string of new drill-pipe, which was not worn out, was charged to this drilling also. Due to difficulty of cleaning hard bed-rock, recovery from this last area sampled was 62% of the estimated value.

The hull of this dredge is 100 ft. long, 43 ft. wide, and 9 ft. deep, and the draft is 8 ft. The deck overhangs the hull 4 ft. on either side. The digging ladder carries 72 buckets of 7-cubic-foot capacity each, and is capable of digging 38 ft. below water level, and of reaching a 15-ft. bank above water-line, making a total digging depth of 53 ft. This is driven by a 150-hp. motor, to which the ladder-hoist is connected also, with provision for independent operation. A 25-hp. motor drives an 8-drum winch. Four motors on pumps range in power from 5 to 75-hp. The 32-inch stacker belt is driven by a 30-hp. motor, and the revolving screen by a 25-hp. motor. Power is furnished from lines of the Pacific Gas and Electric Co. at an average cost of 0.89 cent per kwh., and the consumption per cubic yard of gravel dug is 1.65 kwh.

The dredge was operated three shifts per day with a total crew of 15 or 16 men including the superintendent. Wages in the latter part of 1931 varied from $4.25 per day for oilers to $5.50 for winchmen and $6.00 for the blacksmith. Costs per cubic yard were as follows:

> Period covered: Jan. 1, 1929, to Oct. 1, 1931.
> Cubic yards handled: 2,982,204.
> Total operating cost for period: $241,396.29.
> Cost per cubic yard (33 months): $0.0809.

The cost for the first 19 months was $0.0840, whereas the cost for the last seven months has been $0.0652. The principal reasons for the variation in costs were: (1) Certain physical difficulties in dredging during the early period; (2) several extensive repairs necessary in early period; (3) lower commodity prices during latter period. For the year, 1929, the cost per cu. yd. was $0.0797, which was segregated as follows:

Labor	$0.0255
Superintendency	.0033
Supplies	.0325
Power	.0137
Office expense	.0003
Insurance	.0032
Taxes	.0012
	$0.0797

Lorenz Brothers Hydraulic Mine (No. 123) of five claims is now the property of Trinity County Bank of Weaverville. It is in Sec. 18, 19, T. 33 N., R. 9 W. and Sec. 13, T. 33 N., R. 10 W. During the past season, it has been operated under lease by Paul Finney, Almon Stribling, John Fratus and Harvey Finney, on two eight-hour shifts per day. The water-season usually ends in April or May. The deposit is a reconcentration of the older gravels of the Weaverville Basin by the waters of Weaver Creek. The geology of this basin has been described by J. S. Diller in publications of the U. S. Geological Survey listed below. The gravels of the Weaverville Basin and those of the La Grange mine are thought to be Tertiary in age, and to have contributed much of the gold to the gravels of the present streams in the vicinity. At the Lorenz mine, no low ground is available for dumping tailings, and gravel is elevated to the high end of the sluice with one giant, while a second giant is used for stacking tailings.

Bibl: State Mineralogist's Reports XVIII, pp. 97, 207, 257; XIX, p. 139; XX, p. 182; Pre. Rep. 8, p. 17; Bull. 92, pp. 94–95. U. S. Geol. Survey Bull. 430, pp. 51–56; Bull. 470, pp. 16–18; Bull. 540 A, pp. 5–15.

Fig. 16. Lorenz Hydraulic Mine. (See description No. 123.)

Lost Channel, see Russell.

Lucky Spot (No. 124) is a patented quarter section, NE¼ Sec. 9, T. 33 N., R. 2 W., a mile northwest of Ingot, owned by E. C. Frisbie and others of Redding. An ancient channel is buried with a maximum of 100 ft. of overburden, and has been followed entirely through a hill with an incline, 700 ft. in length. Coarse gold has been produced by various lessees in pieces as large as four ounces. The gold is associated with a white, sticky clay, and presents a rather difficult washing problem. The property was not visited during the present survey, and information on it was furnished by Frisbie and by Frank Gelhaus of Redding, who has been figuring on a lease.

Majestic Mines Co., (No. 125) in 1931, held leases and options on the property of *Humboldt Placer Mining Co.*, about 5000 acres, Buckeye property, 1000 acres, and the Pacific property of about 4000 acres in T. 34 N., R. 8 W., T. 34 N., R. 9 W. and T. 35 N., R. 9 W., covering the Buckeye Ridge between Stuart Fork of Trinity River and Rush Creek. Home office of the Majestic company is at 86 Washington St., Boston, Mass. Phineas Hubbard is president, and A. M. Cripps, secretary.

When the property was visited in the summer of 1931, a crew of 25 men was at work under the supervision of W. D. Ball. A tractor and grader were being used to construct a road along the old Buckeye ditch in Sec. 29 and 33, T. 35 N., R. 9 W. so that lumber could be transported for the building of flumes. About two miles of the road had been finished. The Buckeye ditch is a very old ditch, about 30 miles long, on the opposite side of Stuart Fork from the La Grange ditch. Some prospecting was under way in Baker Gulch near the SW. corner of T. 34 N., R. 8 W. in gravel that probably had been reconcentrated by the present stream from the lower grade gravels of the main Buckeye Ridge. Prospecting had been done on the Stuarts Fork side of the ridge during the preceding winter also. A little later in the year all of this work was discontinued. W. D. Ball was reported to have a small crew at work surveying on Buckeye ground in the summer of 1932.

Bibl: (Buckeye and Humboldt) State Mineralogist's Reports XIV, p. 912; XXII, pp. 37-38. U. S. Geol. Survey Bull. 540, p. 18.

Meckel Property (No. 126) on West Weaver Creek, 2½ miles northwest of Weaverville, is owned by A. C. Meckel of Weaverville, and has been leased during the past season to Manuel Maurica and Frank Lopez.

Phillips, see Big Bend.

Pine Tree, see Upham.

Reddings Creek Placer, Ltd. (No. 127), has acquired the Wallace Bros. mine in Sec. 33, T. 32 N., R. 9 W., and has been operating it during the past two seasons. The company has offices in the Finance

FIG. 17.   Reddings Creek Hydraulic Mine.   (See description No. 127.)

Building, San Francisco, and at 139 N. Virginia Street, Reno, Nevada. Sam Garfalo is secretary and F. C. Wilkins is president and general manager.

During the season, 1930-31, an old ditch giving about 100 ft. of head was in use. With this a piece of ground roughly estimated at 900 ft. long, 150 ft. wide and 8 ft. deep, was mined with a total recovery

stated to have been $4,000. Tailings must be stacked with a giant, as
no low ground is available for a dump.

Installation of new equipment was completed in April, 1931. This
included a 20-ft. dam, 2½ miles of flume, 4 ft. wide and 3 ft. deep, a
30-inch syphon, 300 ft. long of 20-inch for 2000 ft., and two 15-inch
lines, each 1000 ft. long to the mine. Available head is 300 ft. This
new equipment is stated to have cost $40,000. Another item was a
tailings stacker consisting of an inclined grizzly. A giant forces the
gravel up this grizzly, and oversize is discharged at the upper end.
Undersize drops between the inclined bars and is treated in an attached
sluice containing riffles. The outfit is pulled ahead with a 30-hp. tractor
every few weeks as required to keep it close to the portion of the gravel
being mined.

The mine has been operated continuously during the past water
season, ending about July 1, 1932. Four giants were in service, but
water supply is sufficient for only two of these at one time. The giants
are No. 3 with 3-inch nozzle and No. 4 with 5- and 6-inch nozzles. One
giant mines the gravel and delivers it to the base of the stacker, a second
drives it up the stacker, while a third is required to stack the fine tailings of 3-inch size and under. Eight men were employed, and piping
was carried on for 14 hours per day.

The deposit consists of stream gravels of Redding Creek. Bedrock
is a sedimentary series of gravels, sandstones and tuffs of Tertiary age
beneath the portion of the deposit mined recently. A little farther
down the stream Horsetown beds (Cretaceous) are found. Some fine
fossil leaf-prints were shown to the writer by D. S. Upham of Douglas
City, who had been employed at the mine, and who found them in the
bedrock. The particular spot where they were found was later covered
with tailings.

Bibl: State Mineralogist's Report XXII, p. 60.

Rising Sun Mine (No. 128) comprises 190 acres of unpatented
mining claims in Sec. 21, 22, 26, 27, T. 34 N., R. 9 W., held by Basil
Froloff of Lewiston. From Redding, it is 48 miles northwest by road,
35 miles State highway, balance good dirt road. The deposit consists
of the stream gravels of Rush Creek and terrace gravels a little higher
in elevation than the present stream. Depths vary from 10 to 40 ft.,
widths from 400 to 1200 ft.; and the length along the channel is 1¾
miles. Overburden on some of the upper terraces reaches a depth of
from 2 to 10 or 15 feet. Bedrock is a decomposed granodiorite in the
lower quarter mile of the property; the balance is cemented gravel
alternating with beds of clay and sandstone with a dip to the northwest
of 30°. Some good timber grows on the property and there is more
on adjoining land.

For thirty years the upper three-quarters of a mile of this property
was worked by various persons by ground sluicing. Froloff then worked
5 acres of this upper portion with a hydraulic giant and a low head of
water. He states that the yield was 35¢ to 45¢ per cubic yard. During
the season 1931–32, Froloff built a new ditch, half a mile long, on the
lower part of the property with a capacity of about 500 miner's inches,
and giving a head on the gravels near the elevation of the present

stream of from 20 to 40 feet. Other equipment includes 600 ft. of 15-inch pipe and two No. 3 giants using 4-, 5-, or 6-inch nozzles. From 2000 cubic yards that he mined with this equipment, where he was unable to reach bedrock, Froloff states that he recovered $160. In another part of the pit, where he reached bedrock, $27 is stated to have been recovered from 30 or 35 yards. This work indicates that there is an old channel roughly paralleling the present one, but breaking into it here and there. Wm. Carr has a property of 60 acres adjoining at the mouth of Baxter and China gulches.

Russell Mine (No. 129) (Lost Channel Mine) lies immediately south of the Clear Creek drainage in the Igo mining district, Sec. 34, T. 31 N., R. 6 W., and Sec. 4, T. 30 N., R. 6 W., 12 miles southwest of Redding, at an elevation of 1100 feet. The deposit is probably a section of an ancient wash of Clear Creek, and has a width of from half a mile to a mile. F. H. Russell of Redding holds 210 acres patented and additional mining locations to make a total of 500 acres. The present drainage of Upper Dry Creek passes directly across the course of the old channel in a southwesterly direction.

Prospecting of the ground by means of 12 shafts and about 38 drill-holes is stated by Russell to have indicated an average value per cubic yard of $10\frac{1}{3}$ cents for an average depth of 45 feet. The depth to bedrock varied from 18 to 80 feet. Bedrock is largely quartz diorite, in some places decomposed; but a sandstone bedrock is found on the southern part of the holdings.

From the old Dunham tunnel, half of a block of ground 200 by 725 ft. has been removed. The other half still stands in the form of pillars; and the workings are largely open, with a height of from 8 to 10 feet. Samples taken from the pillars are stated to indicate that the gravel removed yielded $60,000. Good pannings can be obtained from certain parts of these workings at the present time. During the season when water was available in 1931–32, Russell and one other man drifted on the gravel at bedrock from a point on Dry Creek about half a mile north of the Dunham tunnel. Some coarse gold was produced from this work, nuggets worth $2 and $3 each.

Bibl: State Mineralogist's Reports XIV, p. 791; XVIII, p. 599; XXII, pp. 186–87.

Steiner Flat (No. 130) is a gravel deposit at the junction of Trinity River and Dutton Creek, three miles down the river from Douglas City. W. D. and G. C. Lorenz of Weaverville holds 180 acres in Sec. 35, 36, T. 33 N., R. 10 W. Hydraulicking on a small scale was done here during the season 1931–32 by W. D. Lorenz. Water supply from the present system lasts only a short time after the winter storms are over.

Bibl: State Mineralogist's Report XIV, p. 914.

Trinity Dredging Company (No. 131) has operated a dredge for many years in Sec. 5, 6, 7, and others, T. 33 and 34 N., R. 8 W., four miles north of Lewiston. Miss Mary Smith of Lewiston is president, and Chas. R. Harris is dredge master. The bucket-line of this dredge carries 42 buckets of 11-cubic-foot capacity and 42 links of the same length as a bucket. When the bucket-line is heavily loaded, these links carry about 2 cu. ft. each. Nine buckets are dumped per minute.

Gravel is washed in a trommel with 6- and 8-inch holes. Oversize, up to 4 ft. in diameter, is dumped over the side through chutes. Undersize goes through the holes in the trommel to a sluice 125 ft. long, 4 ft. wide and 2 ft. deep. The lower part of the sluice, 110 ft. long, is carried on a scow to the rear of the dredge. Separate drums are provided on the winch to swing the sluice for proper distribution of tailings, which are deposited so that the surface is left nearly level. Riffles are made from 2- by 3-inch steel angles with 2-in. face up and 3-in. face vertical. Spacing between angles is 2 inches. The tops are protected with manganese-steel castings, one inch thick, made with bars two inches wide alternating with two-inch openings. These castings are in sections 4 ft. square. Some of these assembled riffle sections, 4 ft. square, are placed with the bars lengthwise of the sluice; but most of them are placed crosswise. The hull of the dredge is 110 ft. long, 50 ft. wide and 7 ft. deep, and draws about 5½ ft. of water. It is provided with two steel spuds of 25 tons each.

FIG. 18.   Trinity Dredge.   (See description No. 131.)

List of motors, all taking power from the lines of the Pacific Gas and Electric Co. at 2200 volts follows: 150-hp. digging motor, geared to line with herringbone gears, no belt, 52-hp. winch motor, driving 10-drum winch, 100-hp. on 16-inch centrifugal pump, 50-hp. on 10-in. centrifugal pump, 35-hp. on 6-in. centrifugal pump, 25-hp. on trommel, 25-hp. on compressor and other shop tools, 50-hp. on a shore pump needed at times to keep the pond full of water.

Bibl: State Mineralogist's Reports XIV, p. 919; XVIII, pp. 601, 734–35; XXII, p. 59, 62; Prelim. Rep. No. 8, p. 18; Bull. 36, p. 104; Bull. 92, p. 95.   U. S. G. S. Bull. 540, p. 20.

Trinity Exploration Co., see Hook and Ladder.

Trinity Farm and Cattle Co. (No. 132), c/o Wm. Foster, Trinity Center, owns some 2000 acres of land on the Trinity River near the center of T. 36 N., R. 7 W., much of which is possible dredging ground. Some drilling to test its possibilities for dredging has recently been done, but the job was not completed, and the results from the holes drilled are not known to the writer.

Trinity River Mining Co. (No. 133), 814 First National Bank Building, Oakland, owns 340 acres on Trinity River in Sec. 5, 8, T. 33 N., R. 8 W., two miles north of Lewiston. A tunnel, 1385 ft. in length, was constructed, giving a head of 25 ft., which was utilized in a turbine and centrifugal pump to force water into pipes for hydraulic mining. This work was done some years ago and the property has been idle for years, but has received some attention recently as a possible source of power to work nearby mines.

Bibl: State Mineralogist's Reports X, p. 709; XII, p. 314; XIII, p. 465; XIV, pp. 915–16. U. S. G. S. Bull. 540, p. 20.

Upham Hydraulic Mine (No. 134) (Pine Tree) consists of 80 acres of patented land and 40 acres of placer claims in Sec. 29, 30, T. 32 N., R. 8 W., owned by F. L. Upham and S. E. Upham of Douglas City. According to F. L. Upham, 50 acres of this is gravel, which his operations indicate will yield 20¢ to 30¢ per cubic yard in gold. His operations indicate a depth of gravel of about 24 ft., and some of it is covered with several feet of overburden consisting of angular slide-rock and soil from the hillside above. The deposit is on three terraces or upper channels of the North Fork of Indian Creek, varying from 15 to 50 ft. higher in elevation than the present stream. Upham has worked out 1000 cubic yards from one pit and 8000 cu. yd. from another, with water from a ditch 45 ft. higher in elevation than the second terrace above the present stream. Bedrock is a hard hornblende schist that has a tendency to form natural riffles across the direction of flow of the earlier stream. A quartz vein about five feet wide is exposed in this bedrock in one of the pits, and it is stated to carry a little gold.

Capacity of present siphons across the creek is 700 miner's inches. Other equipment consists of 1200 ft. of pipe varying in size from 11 to 24 inches, three giants, one each No. 1, 3 and 4. Upham states that a better water supply and a better head could be obtained by building 3000 ft. of flume. The head would be 150 ft. and 2000 miner's inches of water would be available for six months of the year.

Vergnes Property (No. 135). Chester and Henry Vergnes of Oak Bottom (Schilling P. O.) hold 1178 acres in Sec. 7 and 17, T. 32 N., R. 6 W., along Clear Creek. Sec. 17 has recently been purchased from the railroad. The creek flows for a distance of a mile through Sec. 7 and three-quarters of a mile through Sec. 17. Part of the land in Sec. 7 is used for ranch purposes, but is underlain by gravels. On Sec. 17 there are bars of gravel along the creek, roughly 50 or 100 ft. wide and a few hundred feet long. Most of these have already been worked by small-scale placer miners, but according to Henry Vergnes, the gravel still contains fine gold and valuable black sands. There are a few old, caved cuts on quartz stringers on this section also.

On Sec. 7, four different placer operations were recently noted in various stages of development as follows: On a part of the section recently purchased by J. J. Hammer, a very small power shovel, gasoline driven, with a sluice attached to the shovel, had recently done some digging in the bed of the creek. Results had been unsatisfactory, and the shovel was idle at time of visit.

On another part of the ground, a meadow, a shaft had been sunk to bedrock, and a bin and washing machinery were being installed by Chester Vergnes and F. E. Brewster. A 6-inch centrifugal pump was in place at the collar of the shaft, a bin for the gravel, and washing machinery consisting of vibrating screens, sluice, revolving drum with perforations for black sands, and a belt concentrator for treating these. Gravel is to be dumped from cars at the bottom of the shaft into a boot, from which a bucket-elevator is to raise it to the bin. Bucket-elevator had not yet been installed at time of visit.

On another part of the property, Art Affleck, Harry Crites and Roy Kerr had started to mine gravel with drag-line excavators operated by two Fordson tractors. Washing was to be done with water pumped from the creek. At time of visit, repairs were being made to pumps.

On 61 acres of this property purchased from Vergnes Bros. by J. J. Hammer, at the confluence of Boulder Cr. and Clear Cr., Hammer is working the bed of Clear Cr. by hand. He states that it has already been worked, and that what he recovers is mostly surface-wash that has been brought down by recent flood waters. He also occasionally finds gold in crevices in the bedrock.

Wallace Bros., see Reddings Creek Placer, Ltd.

GEOLOGIC BRANCH

CURRENT NOTES

OLAF P. JENKINS, Chief Geologist

Since MINING IN CALIFORNIA is now including the publishing of reports on geology, which represent activity on the part of the Geologic Branch, the name of the quarterly has seemed inadequate. Therefore, the name has been changed to the more appropriate CALIFORNIA JOURNAL OF MINES AND GEOLOGY.

Since the greatest activity in mineral production today lies in gold mining, the Division of Mines has undertaken an intensive study of its recent development. In this issue of the quarterly chapter, the gold deposits of the Weaverville and Redding quadrangles are described in detail by Chas. V. Averill, District Engineer, in conjunction with a geological study made by N. E. A. Hinds, who has for several years been mapping this region, especially in the Weaverville quadrangle. The geologic map presented at this time, though not the final product, shows clearly the areal extent of the formations in both the Weaverville and Redding quadrangles, and also in a part of the Red Bluff quadrangle. In compliance to a special request of the Division, Dr. Hinds has prepared an advance report describing the formations occurring in the area. This map, together with the recently published geologic maps of the adjoining Shasta quadrangle to the north, and the Sierra Nevada region to the southeast, presents a broad picture of the great gold-mining regions of California.

The northwesternmost section of the State, Del Norte and Siskiyou counties, have been heretofore nearly neglected by geologists, so that the following report by John Maxson on a part of this region fills a long-felt need, especially in that he describes the mineral deposits and their relationship to the geology. His paper is entirely a contribution and it is certainly gratifying to know that he intends to continue his studies in the adjoining, even more inaccessible areas.

That there are numerous ways in which geology can be useful to civil engineering has been demonstrated many times, especially in California. At the suggestion of the Geologic Branch, Douglas Clark has prepared a paper on "Some Applications of Geology to Civil Engineering Projects." This also is entirely a contribution on the part of its author.

Lakes, whether natural or artificial, constitute a very important natural resource to California. Their origin, formation and destruction involve features of considerable geologic significance. The very fundamentals of the science are involved in their study, and no one should be better able to discuss a topic on lakes than William Morris Davis, the veteran physiographic-geologist. "The Lakes of California," a generous contribution of its author, published in this issue of the JOURNAL, grew out of a special study of Clear Lake, and it was the pleasure and satisfaction of the Chief Geologist to have accompanied the author of this report on several extremely instructive excursions. The material presented should be of special interest to students of water-supply, engineering, and physiography, as well as instructive to pleasure seekers of lake resorts.

One of the projects of the Geologic Branch, is the mapping of the geology along the eastern front of the Sierra Nevada, especially in Mono and Madera counties. A by-product of this investigation was the discovery of a deposit of a rare mineral known as *piedmontite*, which is described in detail by Evans B. Mayo, temporary field geologist of the Geologic Branch.

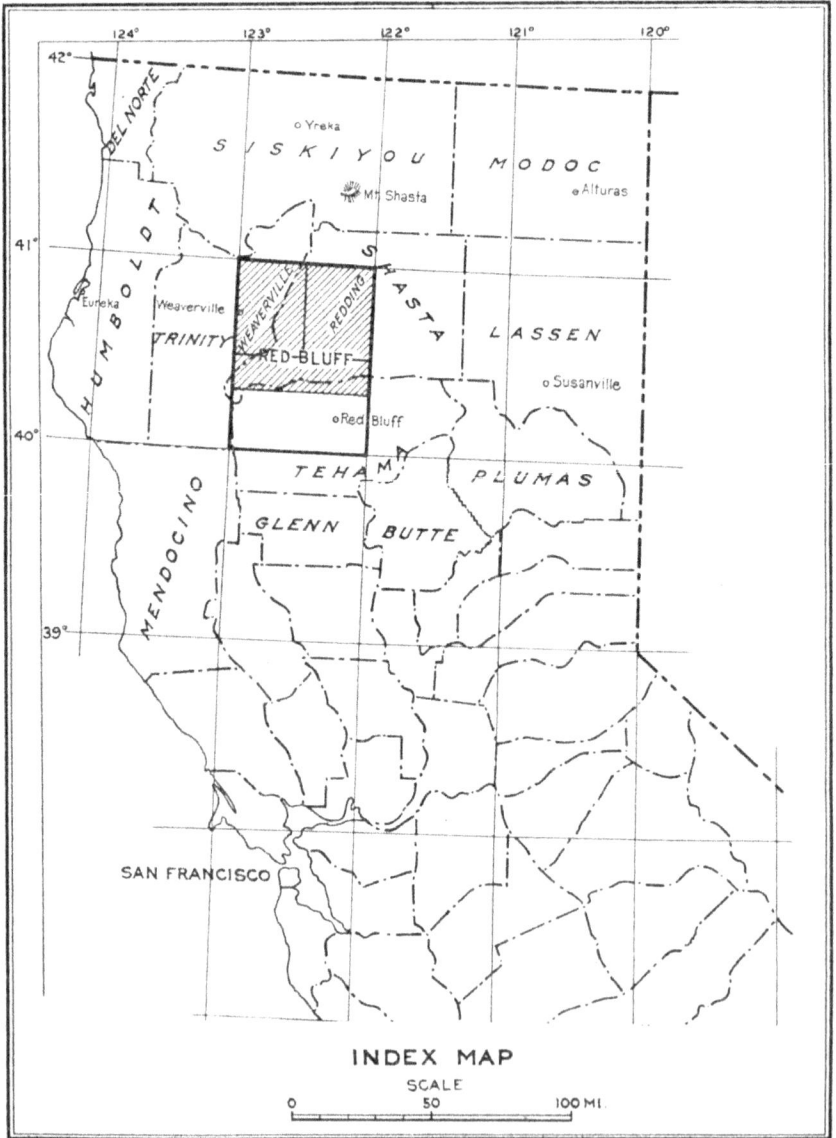

Fig. 1. Index map of northern California showing area covered in report.

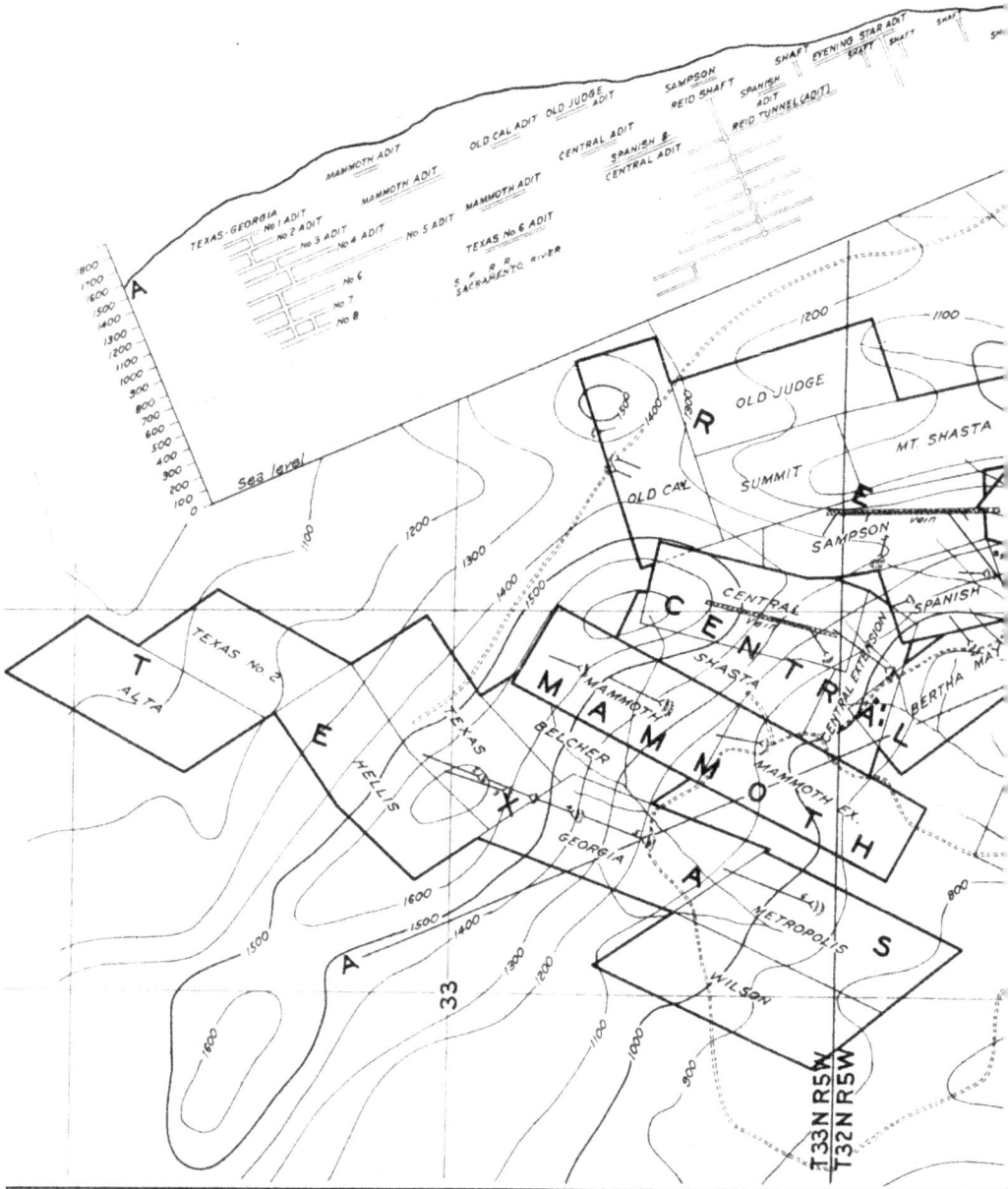

PLATE II

B

WALKER MINE TUNNEL
(ADIT)

MAP OF

OLD DIGGINGS MINING DISTRICT

SHASTA COUNTY, CALIFORNIA.

SCALE

0   600   1200   1800   2400 FT

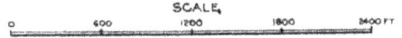

*Accompanying Report on Economic Geology
of Redding & Weaverville Quadrangles.*
by
Chas. V. Averill

DIVISION OF MINES
WALTER W. BRADLEY
STATE MINERALOGIST

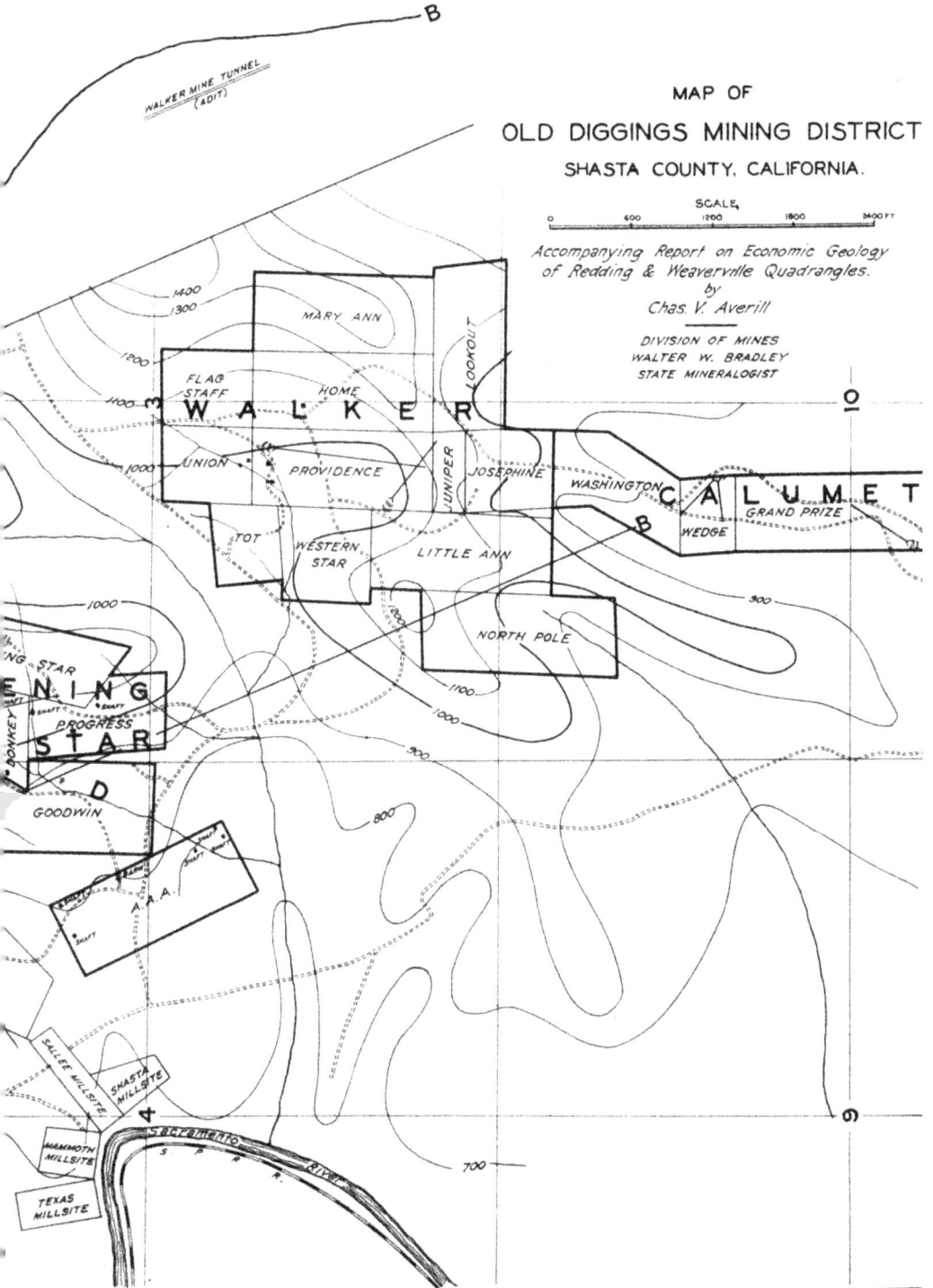

1400
1300
1200

MARY ANN

LOOKOUT

FLAG
STAFF   HOME

1100

WALKER

CALUMET

1000

UNION   PROVIDENCE   JUNIPER   JOSEPHINE   WASHINGTON   GRAND PRIZE

WEDGE

TOT   WESTERN
STAR   LITTLE ANN

B

900

NG STAR

1000

NORTH POLE

SHAFT

ENING
STAR   SHAFT

PROGRESS

800

1100

1000

900

D

GOODWIN

SHAFT   SHAFT

A.A.A.

SHAFT

SALLEE MILLSITE

SHASTA
MILLSITE

4

SACRAMENTO

MAMMOTH
MILLSITE

S   P   R.   R.   RIVER

700

TEXAS
MILLSITE

GEOLOGIC FORMATIONS OF THE REDDING-WEAVERVILLE DISTRICTS, NORTHERN CALIFORNIA

By Norman E. A. Hinds *

	Page
INTRODUCTION	77
General Statement	78
The Geologic Formations	79
Published Reports	79
The Geologic Map	80
Correlation Chart	80
DESCRIPTION OF THE GEOLOGIC FORMATIONS	81
THE SUBJACENT FORMATIONS	81
Pre-Middle Devonian (Probably pre-Silurian)	81
The Siskiyou Terrane	81
Abrams schist	81
Salmon schist	82
Pre-Kennett middle Devonian	85
Chanchelulla formation	85
Copley meta-andesite	86
Devonian	89
Kennett formation	89
Carboniferous. issississippian	91
Bragdon formation	91
Bass Mountain basalt (Bass Mountain diabase of Diller)	91
Baird formation	92
Permain	93
McCloud limestone	93
Nosoni formation	94
Triassic	96
Dekkas andesite	96
Pit formation	98
Hosselkus limestone	99
Brock shale	100
Jurassic. (Pre-Nevadan)	100
Modin formation	100
Bagley andesite	102
Potem formation	102
Late Jurassic. (Nevadan) intrusives	104
Plutonic bodies	104
Trinity Alps stock-boss cluster	104
Shasta Bally batholith	105
Mule Mountain stock	105
Pit River stock	106
Redding quartz augite diorite plutonic dike	106
Hypabyssal intrusives	107
Chonoliths of soda granite porphyry	107
Dikes, sheets, and other small bodies	109
SUPERJACENT SERIES	110
Post-Nevadan Upper Jurassic	110
Knoxville formation (?)	110
Cretaceous	
Shasta series	112
Chico formation	113
Tertiary	114
Eocene	114
Montgomery Creek formation	114
Weaverville formation	115
Pliocene	116
Tuscan formation	116
Tehama formation	117
Tertiary and Pleistocene lavas	119
Quaternary	120
Pleistocene gravels	120
Red Bluff gravels of the Redding Quadrangle	120
Klammath Gravels of the Weaverville Quadrangle	120
Recent alluvium	122

INTRODUCTION

General Statement

At the request of the State Mineralogist and the Chief Geologist of the Division of Mines, the writer has submitted the following report

* Associate Professor of Geology, University of California.

describing briefly the geological formations represented on the accompanying map (Plate 1) which includes the Redding and Weaverville quadrangles and the northwestern part of the Red Bluff quadrangle of northern California. The recently developed renewal of activity in gold mining makes such a statement especially desirable at the present time. A report on the mines of the district by C. V. Averill, District Engineer of the Division of Mines, is published with this paper.

As shown on the accompanying index map (Fig. 1), the quadrangles above mentioned lie in Shasta and Trinity counties of northern California and include a section of the southern Klamath Mountains, adjacent parts of the Cascade lava beds, and the northern end of the Sacramento Valley. The Redding district comprises the southeastern section of the Klamath ranges which are bordered on the east and southeast by the lava covered area of northeastern Shasta and Modoc counties generally included in the Cascade province and on the south by the northern end of Sacramento Valley. The Weaverville quadrangle lies wholly within the southern Klamath Mountains. The mapped section of the Red Bluff quadrangle includes the southern foothills of the Klamath ranges and adjacent part of the northern Sacramento Valley.

This report is essentially a field handbook dealing primarily with the lithologic character and distribution of the various geological formations exposed in the areas mentioned and is in no sense a complete geological report. The petrographic characters of the principal igneous rocks are briefly noted, but descriptions of the many variants to be found in the various igneous bodies or formations lie beyond the scope of this paper. The petrography of the sediments and most of the metamorphic rocks is omitted. In like manner only the major structural features are considered. For further information regarding the geology and structure of the Redding quadrangle, the reader is referred to the Redding Folio, by J. S. Diller, published in 1906 as No. 138 of the United States Geological Survey, Atlas of the United States. A number of short papers dealing with the geology and ore deposits of the Redding and Weaverville districts are listed in the accompanying bibliography. The writer plans to publish a preliminary report on the geology of the Weaverville district in a succeeding number of the CALIFORNIA JOURNAL OF MINES AND GEOLOGY and at some later date a still more detailed report.

The Geologic Formations

A tabular summary of the geological formations exposed in the three principal sections of the area mapped is given below:

	Redding quadrangle	*Weaverville quadrangle*	*Northwestern part of Red Bluff quadrangle*
Recent	Alluvium	Alluvium	Alluvium
Pleistocene	Red Bluff gravels Lava flows	Klamath gravels ----------------	Klamath gravels ----------------
Pliocene	Lava flows Tehama and Tuscan formations	---------------- ----------------	---------------- ----------------
Eocene	Montgomery Creek formation	Weaverville formation	----------------
Cretaceous	Chico formation ----------------	---------------- Shasta series	---------------- Shasta series
Jurassic	---------------- Late Jurassic intrusive igneous rocks Potem formation Bagley andesite Modin formation	---------------- Late Jurassic intrusive igneous rocks ---------------- ---------------- ----------------	Knoxville formation(?) Late Jurassic intrusive igneous rocks ---------------- ---------------- ----------------
Triassic	Brock shale Hosselkus limestone Pit formation Dekkas andesite	---------------- ---------------- ---------------- ----------------	---------------- ---------------- ---------------- ----------------
Permian	Nosoni formation Dekkas andesite	---------------- ----------------	---------------- ----------------
Carboniferous Mississippian	Baird formation Bragdon formation Bass Mountain basalt	---------------- Bragdon formation ----------------	---------------- Bragdon formation ----------------
Devonian	Kennett formation	----------------	----------------
Pre-Middle Devonian	Copley meta-andesite ---------------- ---------------- ----------------	Copley meta-andesite Chanchelulla formation Salmon schist Abrams schist	Copley meta-andesite Chanchelulla formation Salmon schist Abrams schist
Age unknown	Serpentine	Serpentine	Serpentine
Late Paleozoic	----------------	Intrusive igneous rocks	Intrusive igneous rocks

Published Reports

The chief papers describing the geological formations of the mapped area are listed below:

Diller, J. S., The Redding Folio, U. S. Geol. Surv., No. 138, 1906.
    The Auriferous Gravels of the Trinity River Basin, California, U. S. Geol. Surv., Bull. 470, pp. 11–29, 1911.
    Klamath Mountain Section, Amer. Jour. Sci., 4th series, vol. 15, pp. 342–362, 1903.
    Auriferous gravels of the Weaverville quadrangle, California, U. S. Geol. Surv., Bull. 540, pp. 11–21, 1914.
Diller, J. S., and Schuchert, C., Notes on Some Localities of Mesozoic and Paleozoic in Shasta County, California, Amer. Jour. Sci., vol. 47, p. 416, 1894.
Ferguson, H. S., Gold Lodes of the Weaverville quadrangle, California, U. S. Geol. Surv., Bull. 540, pp. 22–79, 1910.
Graton, L. C., The Copper Deposits of Shasta County, California, U. S. Geol. Surv., Bull. 430, pp. 71–111, 1910.
Hershey, O. H., The Metamorphic Formations of Northwestern California, Amer. Geol., Vol. 27, pp. 226–230, 1901.

Hinds, N. E. A., Paleozoic Eruptive Rocks of the Southern Klamath Mountains, California, Univ. of Calif. Publ., Bull. Dept. Geol. Sci., vol. 20, pp. 375–410, 1932.
Smith, J. P., The Metamorphic Series of Shasta County, California, Jour. Geol., vol. 2, pp. 588–612, 1894.

The Geologic Map

The geologic map which this report describes is a composite of the Redding and Weaverville quadrangles and northwestern part of the Red Bluff quadrangle. The Weaverville area was mapped by the writer between 1925 and 1930. A sketch map of a part of the quadrangle, accompanies a report on the gold lodes by H. G. Ferguson[1] published in 1914. In the northwestern part of the Red Bluff quadrangle the contacts of the formations of the Subjacent Series (formations involved in the late Jurassic deformation) were mapped by the writer, those of the superjacent Cretaceous strata (not affected by the late Jurassic deformation) by R. Dana Russell, the writer's associate in the Klamath Mountains investigation from 1928–30. The contacts in this area were mapped on township plots since the topographic sheet long ago published by the U. S. Geological Survey on the scale of 1:250,000 is highly inaccurate and of little use in the field. The township plots are little better; nevertheless it seemed advisable to extend the Weaverville contacts southward to the margin of the superjacent Cretaceous blanket. The area mapped in the Red Bluff quadrangle for the most part lies north of the highway extending from Redding to Platina or Noble Station. The contacts on the Redding section are taken with some modifications from the map by J. S. Diller[2]. The stratigraphic determinations and the nature of certain igneous bodies are based on the most recent geologic and paleontologic work done in the area and differ in a number of respects from Diller's findings, as the following statement indicates:

CORRELATION CHART

Redding Folio by J. S. Diller U. S. Geol. Surv., Atlas Folio 138, 1906	Revised map of Redding quadrangle by N. E. A. Hinds and R. Dana Russell
Pre-Kennett Middle Devonian Balaklala rhyolite	Late Jurassic Chonolith of granite (alaskite) porphyry
Carboniferous Pennsylvanian McCloud limestone	Permian McCloud limestone (lowest Permian)
Nosoni formation	Nosoni formation (upper Lower or lower Middle Permian)
Triassic Pre-Hosselkus Upper Triassic Bully Hill rhyolite	Late Jurassic Chonolith of granite porphyry
Miocene Ione formation	{ Eocene In part Montgomery Creek formation Pliocene In part Tuscan formation
Pliocene Tuscan formation	Pliocene { Tuscan formation { Tehama formation
Pleistocene Red Bluff gravels	Pleistocene Part of Red Bluff as mapped by Diller is Tehama formation

[1] Ferguson, H. G., Gold Lodes of the Weaverville Quadrangle, California U. S. Geol. Surv. Bull. 540, pp. 22-79, 1914.
[2] Diller, J. S., The Redding Folio: Geologic Atlas of the United States Folio 138, 1906.

DESCRIPTION OF THE GEOLOGIC FORMATIONS

The descriptions of the Salmon and Chanchelulla formations, the Copley meta-andesite, the Bass Mountain diabase, the Baird tuffs, and the Nosoni formation are taken with modifications and additions from an earlier paper by the writer [3].

THE SUBJACENT FORMATIONS

Pre-Middle Devonian. (Probably pre-Silurian.)

The Siskiyou Terrane.

Abrams schist.

The Abrams schist was originally described by Hershey [4], during his study in the Klamath and Siskiyou ranges and was named from exposures near Abrams Post Office on Coffee Creek in northern Trinity County. In the western part of the Weaverville quadrangle and the adjacent section of the Red Bluff quadrangle this formation occupies large areas. The principal exposures extend from the Cretaceous overlap northward almost to the town of Weaverville where they are buried by Tertiary beds. A second, small area lying east of and closely related to the first, extends from Mikes Peak in the south central part of the Weaverville quadrangle southward to the north slope of Bully Choop Mountain. A third area occupies the southern slope of Joes Peak just to the east of the area last described. A fourth area of considerable extent is located along the eastern side of the canyon of the Stuarts Fork of the Trinity River from near the mouth of Van Matre Creek northward into the adjacent Shasta quadrangle. In all of these areas the strata are dipping steeply, and in the first mentioned area at least are isoclinally folded so that repetition of the beds occurs.

The Abrams formation consists for the most part of highly micaceous, white, pale gray, or dark gray schists, generally pink or red on weathered surfaces, and composed of colorless or black mica and varying amounts of quartz. In places the mica is green and gives to the rocks a blue-green or dark green color. These schists apparently were derived from clays and shales, much of which contained more or less quartz, and from shaly sandstones. In places the schists grade into micaceous quartz schist with a decrease in the proportion of mica and increase in proportion of quartz. These rocks in turn are transitional into micaceous quartzites and pure quartzites, which were originally slightly impure sandstones and almost pure quartz sandstones respectively. The volume of micaceous quartz schists, micaceous quartzites, and quartzites is small compared to that of the dominant mica schists. A few beds of highly metamorphosed conglomerate and numerous lenses of coarsely crystalline limestone and more finely crystalline marble are present. Along the Stuarts Fork of the Trinity River are considerable thicknesses of rapidly alternating layers of quartzite and mica schist apparently derived from alternating thin beds of chert and clay. Hornblende schists in zones ranging from a few inches to several hundred feet in thickness are present at many horizons in the Abrams;

[3] Hinds, N. E. A., Paleozoic Eruptive Rocks of the Southern Klamath Mountains: Univ. Calif. Publ., Bull. Dept. Geol. Sci., vol. 20, pp. 375-410, 1932.
[4] Hershey, O. H., Metamorphic Formations of Northwestern California, Am. Geol. vol. 27, pp. 226-230, 1901.

the contacts of some of the zones are concordant while others transgress the original bedding of Abrams sediments. These hornblende schists probably represent intrusive phases of the Salmon volcanic activity which followed deposition of the Abrams sediments, though of course some may represent contemporaneous flow of basalt or andesite.

In the great area of the southern Weaverville quadrangle the reddish micaceous schists constitute the great bulk of the exposures; in the area along the Stuarts Fork of the Trinity River the section is considerably more varied, and, while mica schist is the dominant rock, micaceous quartz schist, quartzite, and the rapidly alternating series of quartzites and mica schists are an important element.

The thickness of the formation at the type locality on Coffee Creek in the Shasta quadrangle is estimated by Hershey to be 1000 feet. On the Stuarts Fork of the Trinity River in the Weaverville quadrangle, an incomplete section measures about 2500 feet. In the large area in the southern portion of the Weaverville quadrangle the thickness probably is more than 5000 feet though accurate determinations are uncertain because of the isoclinal folding.

The Abrams is the oldest sequence in the Weaverville district, and, so far as is known, in the whole of the Klamath and Siskiyou mountains. In the Weaverville area the lowermost beds are either in contact with intrusive granites of late Jurassic age or with intrusive serpentines whose exact age has not been determined. Farther to the north the relationships at the lower contact have not been described except for the statement that no older formation has been recognized. Above an erosional contact separates the Abrams from the overlying Salmon schist which is a series of highly metamorphosed basic igneous rocks.

Wherever the writer has seen the Abrams formation and in other places where it has been described, the rocks are thoroughly recrystalized so that none of the original sedimentary types remain. The schistosity in general parallels the original bedding of the sediments so that the initial metamorphism was accomplished without much deformation and probably was of the load type. Remetamorphism has taken place in certain localities as a result of later deformation or as the result of the intrusion of various types of plutonic rocks. One of the most striking zones of remetamorphism is to be found adjacent to a great stock of granodiorite which intrudes the Abrams exposed along the Stuarts Fork of the Trinity River.

Salmon Schist.

Overlying the Abrams schist with erosional unconformity is the Salmon hornblende and chlorite schist also originally described by Hershey [5] from exposures along the Salmon River. Hershey called the formation the Salmon hornblende schist, but in the Weaverville quadrangle there are notable volumes of chlorite schist as well. Hershey considered that the Salmon schist was of sedimentary origin though he later accepted a suggestion made by A. C. Lawson [6] that it probably was a series of fine water-laid volcanic ash. The igneous origin of the formation except for a few sedimentary interbeds have been recently demonstrated by the writer.[7] Two principal areas of the

[5] Hershey, O. H., *op. cit.*
[6] Hershey, O. H., Rocks of Southern California, Am. Geol., vol. 29, p. 272, 1902.
[7] Hinds, N. E. A., *op. cit.*, pp. 388–390.

GEOLOGIC FORMATIONS OF THE REDDING-WEAVERVILLE DISTRICTS   83

Salmon schist occur in the southern part of the Weaverville quadrangle infolded with the Abrams schist; one extends from Hazel and O'Connell Gulch southward virtually to the boundary of the quadrangle and the second is continuous from Indian Creek to the Cretaceous overlap in the north part of the Red Bluff quadrangle. A third area runs through the hilly country about two miles south of Weaverville. The main exposures extend northward from the northern margin of the Tertiary deposits in the Weaverville basin and north of Van Matre Creek are divided into two parts separated by the large tongue of the Abrams schist running along the eastern wall of the canyon of the Stuarts Fork of the Trinity River. The main portion of this area of the Salmon schist lies on the western side of the Stuarts Fork and extends into the Big Bar quadrangle which joins the Weaverville district on the west. A short tongue continues along the Stuarts Fork principally on the western side to a point between Bear Gulch and Devils Canyon. A much narrower belt lies to the east of the Abrams schist area and extends past the north border of the Weaverville quadrangle into the Shasta district. The strata are dipping at high angles in all of the areas mentioned. The greatest thickness exposed is in the area west of Stuarts Fork and probably exceeds 5000 feet though the determinations are uncertain because of complications by folding and the lack of identifiable structures through most of the thickness.

As previously noted the principal rocks of the Salmon formation are schistose and much less common gneissoid rocks derived from basic (probably basaltic and andesitic) volcanics. While various petrographic types are represented, great thicknesses and extensive areas are homogeneous in appearance and in mineralogic composition. One of the most common varieties is a grayish green, olive-green or grass-green, fine-grained, finely foliated hornblende schist generally having a silky sheen or luster. Under the hand lens acicular green hornblende, flakes or chlorite, and small lenses composed of quartz and feldspar are visible. A second exceedingly common type is coarser textured and generally dark green in color. Prismatic, highly lustrous, dark green hornblendes, chlorites, and somewhat larger lenses of quartz and feldspar are visible to the eye. Certain rocks show large hornblende individuals which may reach one centimeter in length. Coarsely banded gneissoid phases are common in certain localities. In the region about the granitoid intrusions in the mountains of western Trinity County, *lit-par-lit* injection of granitic magma into the hornblendic rocks has produced coarse gneisses. The hornblende has been recrystallized and many crystals measure from $\frac{1}{4}$ to $\frac{1}{2}$ inch in length.

The microscope shows that acicular or prismatic crystals of green hornblende compose from 60 to 90% of the hornblende schist. Plagioclase (albite to oligoclase) and quartz are the other principal constituents; the percentage of the two minerals varies, though plagioclase generally exceeds quartz in amount. In some sections quartz is scarce or absent, while plagioclase is always present. Some of the quartz-feldspar lenses are abundantly penetrated by needles of hornblende; others show only a few needles or else are quite free from them. Occasional grains of quartz and feldspar and aggregates of these minerals are present in the hornblendic bands. Epidote and ziosite are rather common associates of hornblende and chloritized biotite generally is present. Minor constituents present in varying amounts are apatite, ilmenite, and magnetite. Some sections contain practically no iron ore minerals.

The hornblende paragneisses are similar mineralogically to the schists except for their coarser foliation and the greater abundance of quartz and plagioclase.

The chlorite schists generally are deep bottle green in color though certain types exhibit a grayish green or olivine-green hue; all have a prominent silky sheen. The schists are fine-grained, highly foliated, and not infrequently rather complexly contorted. The chief visible mineral is chlorite; under a hand lens small amounts of quartz and feldspar are visible in many hand specimens. Specially fine exposures of the chlorite schist may be found in the Trinity Alps along Van Matre, Owen, Boulder, and other creeks flowing from the west into the Stuarts Fork of the Trinity River, and also west of the Middle Fork of Cottonwood Creek in the Red Bluff quadrangle.

The mineralogy of the rocks indicates that they were derived from basic lavas and probably in part from basic pyroclastics which appear to have been erupted upon a rather gently undulating plain eroded into the Abrams sediments. Apparently a great thickness of these volcanics was piled up to form a volcanic plateau of huge area, for the Salmon schist has been recognized from the Klamath Mountains of Tehama County south of the area mapped northward through the northern Klamath and Siskiyou ranges of California and Oregon, a distance of about 200 miles; regarding the east-west extent little is known. The presence of quartzites, micaceous quartz schist, and mica schists (metamorphosed sandstone, impure sandstones, and shales) at various horizons indicates that the volcanic action was interrupted from time to time and that sedimentation went on locally over the volcanic surface. These sedimentary beds were later buried by renewal of volcanic eruptions.

In a paper previously referred to,[8] the writer considers the Abrams and the Salmon formation member of a single terrane for which is suggested the name Siskiyou from the wide exposure of these beds in the Siskiyou Mountains where they were first described. These formations, though of different ages, form a distinct stratigraphic group since they are much more highly metamorphosed than any of the later formations so far known from this region and their degree of recrystallization is similar. Their geologic age is uncertain. Pebbles and boulders of rocks similar to the Abrams and Salmon schists are included in the overlying Chanchelulla conglomerates and the difference in degree of metamorphism between the Chanchelulla and the strata of the Siskiyou terrane everywhere is striking. Therefore, a long time of the earliest Paleozoic intervened between the deposition of the Abrams sediments and the eruption of the Salmon volcanics and the invasion of the middle Devonian Kennett sea, but the periods involved can not as yet be stated. The writer believes that the Siskiyou formations are definitely pre-Silurian[9] and probably pre-Cambrian though as yet proof of this is lacking. Some time elapsed between the deposition of the Abrams sediments and the eruption of the Salmon lavas, but there is yet no basis for determining their relative ages.

[8] Hinds, N. E. A., *op. cit.*, p. 387.
[9] Hinds, N. E. A., *op. cit.*, p. 391.

Pre-Kennett Middle Devonian

Chanchelulla formation

In the western Klamath, a very thick series of sediments of undetermined age unconformably overlies the strata of the Siskiyou terrane with moderate angular and distinct erosional unconformity and forms a broad belt running from the edge of the Cretaceous overlap on the subjacent complex in the Red Bluff quadrangle northwestward through the Weaverville and Big Bar quadrangles, and perhaps beyond. The full extent of the formation has not been ascertained. The base of the sequence is composed largely of thinly bedded gray or much less commonly red and green cherts which have suffered extensive recrystallization to massive or platy quartzites. The original stratification of the cherts shows in most exposures. Interbedded with the cherts are micaceous, graphitic, and calcareous schists, quartzite, metaconglomerate, marble, some of which is graphitic. Toward the top of the formation, the proportion of mica schist, partially metamorphosed sandstone and conglomerate, and limestone is considerably greater though chert continues as an important element. The total thickness exceeds 5000 feet. The strata have been strongly folded and moderately metamorphosed, though the recrystallization has been much less intense than in the Siskiyou terrane.

This formation has not been recognized in previous discussions of the stratigraphy of the southern Klamath. The writer therefore proposes for it the name *Chanchelulla* from the excellent exposures on the slopes of a mountain of that name in the northwest corner of the Red Bluff quadrangle. A more detailed description of the lithology and stratigraphy of this formation will be published in the preliminary report on the Weaverville quadrangle now in preparation.

After the sediments had been deposited, they were flooded by great volumes of basic magma which solidified chiefly as sills of andesite and andesite porphyry, some of which are 200 to 500 feet in thickness and have great areal extents. Dikes and a few irregular bodies also are present. Since their intrusion, these rocks have been metamorphosed to greenstone or chlorite schists. The date of this volcanic episode is uncertain. Mineralogically the meta-andesites very closely resemble the Copley volcanics subsequently to be described and their degree of metamorphism is similar to that of the Copley rocks. The suggestion is strong from these facts and from field relations observed in a few places that the basic rocks in the Chancelulla formation are intrusive phases of the Copley meta-andesite, but this needs confirmation by further studies.

The age of the Chanchelulla sediments has not yet been determined. Radiolarian remains are present in the cherts but have been more or less damaged by recrystallization of these rocks. The limestones in all cases examined have been completely recrystallized and fossils, if originally present, have been destroyed. None have been found in other types of sediments. The formation overlies the beds of the Siskiyou terrane with pronounced erosional and slight angular discordance. The presence in certain of the Chanchelulla conglomerates of metamorphic rocks apparently derived from the Siskiyou beds indicates that the time interval separating them is long.

The Chanchelulla in turn is overlain by remnants of the once great Copley lava plateau which is of pre-middle Devonian age. Thus Chanchelulla is a pre-Copley formation belong to some Eo-Paleozoic period.

The Chanchelulla meta-andesites bear close megascopic resemblance to certain of the Copley volcanics subsequently to be described, though they more commonly possess schistose structures than do the Copley rocks. The meta-andesites are grayish green, olive green, or dark green in color and in texture range from medium to aphanitic. Porphyritic varieties with plagioclase and augite as the visible phenocrystic minerals are dominant. Along shear zones, dark green, fine-grained, highly foliated chlorite schist has been developed.

Microscopic examination shows a wide variety of textures and structures. The original rocks were augite and hypersthene andesites and andesite porphyries composed of medium or calcic andesine, augite or hypersthene, and accessory magnetite and apatite. Metamorphism has converted the rocks to massive or schistose greenstones. In the massive types, relic grains of the original minerals are rather plentiful so that the initial characters can be determined. The feldspars have partly gone over to sericite, calcite, and kaolin; the ferromagnesians to chlorite, epidote, zoisite, calcite, and in a few cases to green hornblende. Further recrystallization has more or less completely destroyed the original minerals and chlorite schists have been produced. In most sections examined, the schistosity is only rudely developed, but where metamorphism has been specially intense, as along shear zones, highly foliated chlorite schists are present.

Copley Meta-Andesite

The description of the Copley meta-andesite is taken with modifications from an earlier paper by the writer.[10]

Widely distributed though the Klamath and Siskiyou mountains are remnants of a great volcanic series called by Diller the Copley meta-andesite. The principal areas of this formation in the region mapped lie in the western half of the Redding quadrangle and the eastern half of the Weaverville quadrangle along the boundary between the two, but other areas have been mapped or reported farther to the west in the Red Bluff, Weaverville, and Big Bar quadrangles, and to the north in the Shasta quadrangle. Objection may possibly be raised to the correlation of certain isolated patches in the western and northern Klamath-Siskiyou region with the Copley, but their stratigraphic relations and lithologic characters so closely resemble those of the main exposures that little doubt can be entertained regarding the validity of such correlation.

The Copley formation consists of fissure flows of pyroxene andesite and less commonly of basalt with which are interbedded large volumes of coarse and fine pyroclastics. Dikes and sills, which appear to have been erupted during this volcanic epoch, are present. The complex at some time prior to the middle Devonian formed a lava plateau of considerable extent. The area over which exposures are present is about 5500 square miles but this doubtless represents only a fraction of the original extent. Subsequent deformation and metamorphism have greatly altered the structure of the plateau and the character of its

---

[10] Hinds, N. E. A., *Op. cit.*, pp. 393-400.

rocks and long continued erosion has stripped away large areas and has reduced the thickness of the remaining parts. The present thickness in most places is difficult to determine; in the Redding district Diller estimated about 1000 feet, but Louderback [11] has recently described part of a much thicker section near Kennett on the Sacramento River. The writer has measured two sections in the Weaverville quadrangle which exceed Diller's figures, but in neither case is the total thickness exposed. One on Stacey Creek measures more than 1200 feet, the second on Shirttail Peak about 1500 feet. Many good exposures are unsuitable for measurement since metamorphism has destroyed virtually all evidence of structure.

The partial section examined by Louderback was studied in the field and by means of drill cores in connection with surveys for a dam site. A few details are given below:

West side of Sacramento River—elevations measured above the river.

Elevation 400–600 feet—prominent outcrops of agglomerate.

300 feet—fine-grained, somewhat schistose andesite. In part this rock has been converted to chlorite schist.

Government road to 270 feet—dull, ochreous, friable, "shaly" meta-andesites followed by badly weathered amygdaloidal andesite.

145 feet—amygdaloidal andesite.

150–155 feet—chlorite schist.

At and near river—agglomerate composed of andesite fragments.

East side of Sacramento River

Near river—amygdaloidal and porphyritic meta-andesite with bands of chlorite schist. Fragmental types increase in volume to end of measured section at an elevation of 370 feet above river. In this section are dikes of augite diorite and dacite porphyry which apparently are younger than the Copley. The section was measured nearly at right angles to the strike which is approximately parallel to the course of the Sacramento River at this point. The strata dip at angles which do not depart more than 20 degrees from the vertical. The thickness included between the east and west limits is about 2000 feet, but this probably is considerably less than the total thickness of the Copley in the area.

The Copley volcanics everywhere have been metamorphosed to greenstone schists but the intensity of recrystallization varies from place to place. In some areas a considerable proportion of the original minerals of the lavas and pyroclastics can be readily determined. Porphyritic flows, vesicular and amygdaloidal lavas, and brecciated zones of blocky or clinkery lavas are not uncommon. Beds of tuff, breccia, and agglomerate are abundant so that much may be learned from such exposures regarding the plateau and the composition and structure of its rocks. Sills and dikes have been distinguished at a number of localities. More commonly, however, metamorphism has destroyed the initial structures and textures and the exposures are composed of relatively homogeneous, massive or schistose rocks. Detection of the original relations also is impeded by weathering which has produced a reddish or brownish oxidized layer giving a homogeneous appearance even to very large exposures.

Along shear zones and near the contacts with the large subjacent intrusions, the Copley volcanics have been again recrystallized to

[11] Bailey, P., The Development of the Upper Sacramento River. Bull. 13, State of California, Dept. of Public Works, pp. 56–59, 1928.

chlorite and less commonly hornblende schists. The latter in many places are difficult to distinguish from the Salmon schists previously described and a careful study of the stratigraphy or tracing of the schists into the normal Copley rocks is necessary for their correct identification. An especially fine development of such a schistose phase is in the great roof pendant between the Mule Mountain and Shasta Bally quartz diorite intrusions of Jurassic age in the southeastern quarter of the Weaverville quadrangle. Certain geologists have considered this schist a separate formation unit, but to the north and east it grades into the normal Copley and consequently its true relationship can be positively established.

In this zone and others of similar character subjacent metamorphism has been especially intense and has been superimposed on an earlier metamorphism which had converted the basic rock into massive metabasic equivalents.

Along the larger shear zones, belts of chlorite schist are commonly developed. An excellent example may be studied along the fault system at the head of Modesty Gulch about three miles northwest of Oak Bottom on which the Ganim mine is located. Dynamic and thermal metamorphism have converted the massive meta-andesite into talc and chlorite schist which in certain zones are abundantly impregnated with pyrite and with subordinate amounts of chalcopyrite. At the Ganim Mine this sulphuretized zone is well exposed at the surface, while in a tunnel which cuts the fault zone a large body of massive and schistose talc of high grade has been encountered.

Along the splendidly exposed normal fault at the La Grange Mine about five miles west of Weaverville, chlorite schist has been developed from the meta-andesite. Where the rocks are broken by many small faults, alternations of massive andesite and chlorite schist are generally present. In all areas of the Copley which the writer has examined, these schistose zones are abundant.

The most conspicuous fragmental rocks are breccias and agglomerates in which the fragments range from 4 to 8 or 10 inches in diameter, though in many places masses from 2 to 3 feet in diameter are present. Tuffs are common and probably are much more abundant than the exposures indicate since they have been converted into massive metamorphic equivalents and megascopically bear close resemblance to finely granular, massive lavas. All of the tuffs so far observed are lithic crystal types with fragments rarely exceeding one inch in diameter. The coarser pyroclastics generally are recognizable because of the difference in color between the matrix and the fragments or because of differential weathering which has caused the fragments to stand either above or below the general surface of the rock. These effects are conspicuous both on fresh and weathered surfaces. The pyroclastics are dominantly andesitic; basalt, quartz porphyry, dacite, granite, granodiorite, and diorite fragments are present in minor amounts. In certain areas quartz porphyry fragments are rather abundant.

The typical Copley lavas and pyroclastics are yellowish green, grayish green, and dark green in color. The prevalent greenish shade is due to the general development of chlorite from the ferromagnesian minerals. Not infrequently, epidote is so abundant that it imparts a yellowish green hue to the rocks. The weathered surfaces are brown, yellowish brown, reddish brown, or red.   Porphyritic textures are common with augite and plagioclase as the chief phenocrystic minerals. Occasional grains of magnetite also are visible to the eye.   Many of the flows are amygdaloidal and their presence is of great assistance in working out original structures.   Quartz, calcite, and less commonly chlorite are the chief vesicle fillings.   Rather well developed flow structure is present in certain lavas.   Where metamorphism has obliterated original textures and structures, large areas and considerable thicknesses appear as homogeneous, fine- and even-grained rocks in which it is impossible to detect flow surfaces or the bedding planes of pyroclastic deposits.   Slaty cleavage is rather conspicuous in certain of these areas.

In certain areas the stratigraphic relations of the Copley are easily determinable; in others either faults bring the volcanics in contact with the adjacent formations or else the contacts are not exposed. In the western Klamath the volcanics lie with pronounced angular and erosional discordance upon the rocks of the Siskiyou terrane and appear to be similarly related to the Chanchelulla strata.   Farther to the east the base of the Copley is nowhere exposed.   In the Redding district the middle Devonian Kennett formation, or more commonly the Mississippian Bragdon, lies upon the deformed and eroded surface of the Copley.   The exposures of the Kennett-Copley contact are limited but it is clear that the Kennett was laid down upon the eroded surface of the Copley lava plateau; contacts between the Bragdon and Copley formations are more commonly met than those between the Kennett and the Copley.   The basal Bragdon and also the basal Kennett strata in places are composed of detritus apparently derived from the Copley volcanics.   Pebbles of greenstone are found in the conglomerates and certain of the Bragdon sandstones contain large quantities of plagioclase feldspar, chlorite, and ferromagnesian minerals petrographically similar to the components of the volcanics.

No interbedded sedimentary rocks have so far been found in any of the Copley exposures so that interruptions in the volcanic history are not indicated by field evidence so far at hand.   That such interruptions occurred and that sedimentation went on over the temporary surfaces of the plateau are very probable and further study of the areas of the Copley may reveal intercalated beds of sedimentary origin.

From the previous discussion it is evident that positive determination of the age of the Copley at present can not be made.   The formation is separated from the middle Devonian Kennett sedimentation by a period of folding and erosion and hence must considerably antedate the Kennett formation.

Devonian

Kennett formation

The Kennett formation of Middle Devonian age is the oldest fossiliferous sequence in the Klamath Mountains and consequently is of great importance in the determination of the stratigraphy of the region. The formation is made up chiefly of black shales and slates which are in places highly siliceous; thin-bedded sandstone; and lenses of highly fossiliferous, dark gray to bluish gray limestone, most of which has been recrystallized to marble.

The Kennett formation appears only in the Redding division of this map sheet. The best exposures lie on the western side of the Sacramento Canyon to the west and north of the old mining town of Kennett about 16 miles north of Redding. The principal area is located on both sides of Little Backbone Creek and extends eastward across the Sacramento River; here the limestones are thick and extensive and form bold outcrops along the upper slopes of the Sacramento Canyon. Other areas are located (1) on the southern side of Little Backbone Creek, (2) near the summit of Bohemotash Mountain, (3) on the eastern slope of O'Brien Mountain, (4) east of Kennett and extending on both sides of Sacramento River, (5) on the south slopes of Bass Mountain and extending out into the adjacent flat, (6) south of the old settlement of Churntown, (7) east of Newtown, (8) about 4 miles west of Redding and immediately south of the bend of the Sacramento River. Other smaller areas also are present. A section of the Kennett along Backbone Creek north of the town of Kennett measured by Stauffer,[12] is given below:

Mississippian   Thickness in feet
    Bragdon formation:
        9. Conglomerate with quartz pebbles and fossiliferous frag-
           ments, these latter often dissolved leaving holes_____ 30

Devonian
    Kennett formation:
        8. Shale, mostly dark, with thin sandy beds_____ 140
        7. Limestone, massive, light gray with a small amount of
           chert. This rock has been quarried to some extent on
           Quarry Ridge. Part of it is filled with corals_____ 100
        6. Sandstone, thin-bedded, and dark gray shale_____ 140
        5. Limestone, dark gray to bluish_____ 10
        4. Shale, cherty, gray_____ 150
        3. Limestone, thin-bedded to massive, often cherty and parts
           of it full of various types of corals. A black chert band at
           the base_____ 200
        2. Shale, siliceous, often sandy and usually very thin-bedded.
           Shales mostly black to gray in color and often partly meta-
           morphosed_____ 65

Balaklala rhyolite:
        1. Reported as rhyolite, but is an intrusive igneous mass
           (alaskite porphyry). A greenish and grayish rock with
           phenocrysts of quartz and feldspar_____

The sections exposed in the other areas are less complete.

The Kennett lies unconformably upon the deformed and eroded Copley volcanics and the basal strata of the Kennett contain rock fragments very evidently derived from the latter rocks. Above, the Bragdon strata of Mississippian age lies unconformably on the Kennett. Before the deposition of the Bragdon there had been very extensive erosion of the country and therefore the Bragdon strata lies upon disconnected erosion remnants of the Kennett beds.

According to Stauffer,[13] an abundant fauna consisting chiefly of corals present for the most part in the limestones are indicative of a Middle Devonian age.

_____
    [12] Stauffer, C. R., The Devonian of California: Univ. Calif. Publ., Bull. Dept. Geol. Sci., vol. 19, pp. 95-96, 1930.
    [13] Stauffer, C. R., op. cit, p. 104.

Carboniferous. Mississippian

Bragdon formation

The Bragdon is the most widely distributed sedimentary formation in the southern Klamath Mountains. The main area occupies the northeastern quarter of the Weaverville district and the northwestern quarter of the Redding district and extends far to the north into the Shasta quadrangle. Smaller exposures are numerous in the western part of the Weaverville district and two exposures, one at Harrison Gulch and the second on Jerusalem Creek, are located in the northwestern part of the Red Bluff quadrangle. The formation is composed chiefly of thin-bedded shales which are commonly slaty; medium to fine-grained, generally dark gray, very firmly cemented sandstone; and conglomerates in which the pebbles are chiefly quartz and chert. The formation probably exceeds 6000 feet in thickness although the actual measurements are difficult to obtain because of complication by folding. The Bragdon unconformably overlies erosion remnants of the Kennett formation but more commonly is either in erosional contact or fault contact with the still older Copley meta-andesite; evidently long emergence accompanied by widespread and deep erosion took place after the close of Kennett submergence.

The great bulk of the Bragdon formation is composed of thinly bedded slates or shales which are generally gray in color or even black where the carbon content is especially high. These shales and slates probably compose more than 90 per cent of the formation. The sandstones generally consist chiefly of chert and quartz grains, but certain sandstones near the base of the formation contain large quantities of debris from the Kennett and Copley formations. In the conglomerates also are abundant fragments coming from both of these underlying series; in places these rocks contain fossiliferous Kennett limestone pebbles. The fossils found in the Bragdon indicate its Mississippian age. The fossils are scarce and many are poorly preserved. Within the Bragdon formation in the Redding quadrangle is the important volcanic sequence called the Bass Mountain basalt which is subsequently described, and besides this there are many horizons of tuffaceous sandstone and slate. Tuffaceous elements are also present in certain of the sandstones.

Bass Mountain basalt (Bass Mountain diabase of Diller).

Diller [14] described the Bass Mountain diabase as a series of flows, tuffs, and volcanic and other sediments forming a part of the Bragdon formation in the Redding quadrangle. The application of the term "diabase" to these surficial eruptives is objectionable in view of the tendency among many geologists to confine the term to intrusives possessing the special texture known as "diabasic"; hence Bass Mountain basalt is used.

There are two principal areas in which the volcanics are exposed: (1) on Bass Mountain and southward to Newtown in the central part of the Redding quadrangle, and (2) discontinuous patches along Middle Salt Creek, a tributary of the McCloud River in the northern part of the same district. At the first locality the volcanics rest unconformably on the Copley meta-andesite and the Kennett formation and are inter-

---

[14] Diller, J. S., *op. cit.*, p. 7.

bedded with and overlain by Bragdon strata. Diller stated that the various associations exhibited by the Salt Creek section indicate "contemporaneous volcanic activity in the later part of the Bragdon epoch" and that the rocks of the Bass Mountain area probably also belong to the same horizon, "though the evidence is not altogether satisfactory. On Bass Mountain it appears that the eruptions were taking place contemporaneously with the deposition of Bragdon sediments, but whether this activity was contemporaneous with that at Salt Creek is not yet determined."

The Bass Mountain lavas are dense, dark gray or greenish gray. medium- to fine-grained rocks in which plagioclase, augite, olivine, chlorite, magnetite, and less commonly hornblende and quartz are visible to the eye or under a hand lens. Porphyritic textures are common with augite and plagioclase as the chief phenocrysts. A few flows are amygdaloidal.

Microscopic examination shows that most of the lavas are holocrystalline, generally porphyritic, and possess either ophitic or intergranular textures. Two sections show a considerable amount of completely devitrified glass. Microphenocrysts of plagioclase (labradorite to bytownite), augite, and in some specimens olivine are set in a groundmass of plagioclase, augite and accessory magnetite and apatite. The phenocrysts are more or less completely altered to calcite, sericite, and kaolin; the groundmass individuals are fresher and from them the composition of the plagioclase is readily determined. Colorless augite is the principal ferromagnesian mineral; both phenocrysts and groundmass individuals are extensively chloritized. Epidote, calcite, and fibrous green hornblende also have been developed from the augite. Where olivine is present, the crystals have been almost completely serpentinized. Certain of the lavas in the Bass Mountain area contain light brown hornblende as well as augite and their plagioclase is calcic andesine.

Fine- to medium-grained lithic tuffs are present in the series and probably exceed the flows in volume. These rocks are dark colored, very firmly lithified and are composed of fragments of basalt like those of the Bass Mountain flows.

Baird formation

Overlying the Bragdon sediments with apparent conformity is the Baird formation containing an abundant fauna which J. P. Smith [15] correlates with the higher divisions of the Lower Carboniferous of the Mississippi Valley and the Lower Carboniferous of the Eureka district, Nevada. The upper part of the formation, which contains most of the fossils, consists of 150 to 200 feet of sandstone and calcareous and siliceous slates most of which are tuffaceous. Below are about 500 feet of tuffs and less abundant tuffaceous sandstones and slates. As originally defined by J. P. Smith, the formation includes about 500 feet of "black metamorphic siliceous shales" below the McCloud limestone, but Diller extended the lower limit to include "the tuffaceous rocks and the adjoining sandstones and shales which overlie the topmost Bragdon conglomerate."

[15] Smith, J. P., Metamorphic Series of Shasta County, California, Jour. Geol., vol. 2, pp. 595-597, 1894.

From the Pit River northward, the Baird forms an irregular belt ranging from one-half to two and one-half miles wide. On the west the strata generally are in contact with the dike of augite diorite previously referred to, which rose along the plane of separation between the Baird and the overlying Permian McCloud limestone. South of the Gray Rocks, the Baird is represented by a narrow belt about one-fourth of a mile wide.

At various localities the tuffs and sandstones are well exposed; the best sections lie immediately north of the fishery and along the Pacific Highway where it starts across the hills from the McCloud into the Little Sacramento Canyon. The tuffs exposed near the fishery are medium to dark gray, well lithified rocks in some of which angular fragments of andesite and rhyolite up to one inch across are abundant. Many of the tuffs are fine-grained, but under the microscope are similar in composition to the coarser varieties.

Farther up the McCloud Canyon along the Salt Creek road, Diller [16] records fine exposures of the tuffs which he describes as made "wholly of volcanic material and crowded with feldspar microlites, but most of it rhyolitic. Some of the fragments are devitrified, showing traces of original perlitic cracks; others contain anhedral quartz embedded in a mosaic of quartz and feldspar."

Permian

McCloud limestone

The McCloud limestone, conformably overlying the Baird formation and in turn overlain by the Nosoni beds along the west side of the McCloud canyon, can be traced more or less continuously from the northern border of the Redding quadrangle for about 25 miles to the south where it disappears near the settlement of Lilienthal. At its southern extremity, the limestone appears as lenses, but a short distance to the north it gradually thickens, becomes continuous, and, north of the Gray Rocks, is a prominent member of the stratigraphic column. The limestone is exceedingly resistant to erosion and forms bold, rugged ridges or mountains wherever it is exposed; along the eastern side of the McCloud River, great, jagged limestone peaks are the most conspicuous feature of the landscape. The McCloud, through most of its thickness, is composed of pale gray to dark gray fine-textured marble; practically none of the original limestone remains. Most of the rock is massive, but finer-stratified zones also are present. Locally, and especially along igneous contacts, the recrystallization has been much coarser and generally the gray color has been bleached out leaving a very pale gray or white rock. Chert layers, lenses, and nodules have been developed apparently as a result of metamorphism by solutions associated with igneous intrusions; these siliceous zones are more resistant to weathering than the limestone and consequently they stand out rather conspicuously from the limestone surfaces. They are also pale brownish or buff in color and contrast with the normal gray of the limestone. At the southernmost exposures near Lilienthal, the McCloud is about 200 feet thick while along the east side of the McCloud Canyon it reaches a maximum of 2000 feet on Horse and Town mountains. The McCloud limestone is so highly fossiliferous that most

---

[16] Diller, J. S., *op. cit.*, p. 3.

weathering surfaces show some representation of the fauna. Beds composed of cup corals, protozoa, and crinoid stems are common. Internal structures have been commonly destroyed by recrystallization. The fauna consists chiefly of corals and protozoa (Fusulina); brachipods and gastropods are fairly abundant. Crinoid stems in great profusion are present but complete individuals are rare. According to Diller [17] the age of the McCloud is Pennsylvanian, but recent work by H. A. Wheeler [18] of Stanford University has shown that the fauna is of Lower Permian age and that the Pennsylvanian is not represented in this region.

While the McCloud was laid down on the Baird strata, it is at present separated from the Baird throughout much of its extent by an enormous dike-like body of quartz augite which apparently came up along the contact of the two formations. The Baird and McCloud are in contact north of Hirz Mountain for about four miles and also south of Gray Rocks for about four miles. As a result of the intrusion of the diorite, the limestone was very greatly shattered and many huge blocks or xenoliths were engulfed in the dike-rock. These show up as isolated masses completely surrounded by the augite diorite along the whole length of the dike. Many of these xenoliths show more intensive recrystallization than the average of the formation. Along the contacts of the igneous body and the limestone, a considerable suite of metamorphic minerals have been developed of which hedenbergite, magnetite, and garnet are the most conspicuous. Small bodies of magnetite or magnetite rock containing various proportions of other minerals, principally garnet, are found along the contact within the limestone or within the igneous rock. Some magnetite was mined from these bodies on the west side of the McCloud canyon near the post office of Baird about one and one-half miles north of the junction of the Pit and McCloud rivers. The limestone has been used for flux at mines of the Bully Hill district farther to the east.

Nosoni formation

Diller [19] described under the name Nosoni formation a group of volcanics and sediments lying between the Pennsylvanian McCloud limestone and the Triassic Dekkas andesite. Fossils, which are exceedingly abundant in certain zones, were believed to indicate a position at the top of the Carboniferous, but later studies have shown that a Permian age is more probable.

The major part of the Nosoni is composed of lithic crystal tuffs and agglomerates made chiefly of basaltic fragments; crystals of plagioclase and femic minerals; flows of andesite and of olivine basalt also are present. Interbedded with the volcanics and dominant in the uppermost part of the formation are sedimentary strata which included dark brown, fossiliferous, shaly limestones which have been subjected to extensive silicification; dark gray to brown, medium grained, shaly, tuffaceous sandstones; and dark gray tuffaceous shales and slates. Fossils are present in the sandstones and shales as well as in the limestone.

[17] Diller, J. S., op. cit., p. 4.
[18] Wheeler, H. A., Private communication.
[19] Diller, J. S., op. cit., p. 5.

The lavas Diller describes as

"dark and compact, with uneven fracture. In most cases they are holocrystalline, decidedly microporphyritic, and made up chiefly of plagioclase, with much augite, less magnetite, and occasional olivine or iddingsite. The tabular feldspars are enclosed in a groundmass having interstitial structure modified by fluidal movement giving the lath-shaped crystals of labradorite a parallel, stream-like arrangement with the granular augite and oxides of iron between them. Augite occurs sparingly among the phenocrysts, but olivine or iddingsite is common though neither appears in the groundmass and both are sometimes entirely absent."

Most of the lavas show considerable alteration with chlorite and calcite as the chief secondary minerals developed. In the few flows which are amygdaloidal, calcite, quartz, and chalcedony are the chief vesicle fillings; chlorite also is present.

The Nosoni forms a belt ranging from three-eighths of a mile in width near the overlap of the Cretaceous and Tertiary formations on the subjacent complex to $3\frac{1}{2}$ miles wide along the northern margin of the Redding quadrangle. Northward in the Shasta area, the formation passes under the Columbia-Cascade lavas. The well exposed section on Horse Mountain east of Baird is about 500 feet thick while on Nawtawakit Mountain 10 miles to the north a maximum of 1200 feet were measured. In these thicker parts the proportion of pyroclastics is greater than to the south. The sediments are chiefly at the top of the formation though sedimentary interbeds are present throughout.

Regarding the relations of the McCloud and Nosoni, Diller [20] writes,

"The Nosoni conformably overlies the McCloud limestone opposite Horse Mountain and farther south, but to the north the accumulation of so great a mass of pyroclastic rocks beneath the main horizon of fossils suggests an unconformity. However, the occurrence near the bottom of the pyroclastics of small lenses of shale bearing essentially the same fossils as those above the close affinity of the fauna with that of the McCloud limestone strengthen the evidence in favor of conformity."

Recent study of the McCloud-Nosoni contact shows the presence of a pronounced erosional discordance which implies withdrawal of the earlier Permian sea and the deep erosion of the limestone surface followed by reinvasion of the ocean in later Permian time and deposition of the Nosoni beds. The disconformity is particularly well shown in the long subsequent valleys developed in the Nosoni south of the Pit River. According to Beede and Kniker,[21] who have studied the protozoan limestones of the McCloud,

"in California, the relation of the Permian and Pennsylvanian section is still uncertain because of the meagerness of the data available due to physical conditions of the deposits. However, the association of *Schwagerina robusta* Meek, *Fusulina cylindrica* and *Fusulina extensa* var. *californica* Staff is a normal association which may be expected and is usually found in the regular *Schwagerina* horizons of other regions. * * * As the matter now stands, the age of the California *Schwagerina* beds is not as definitely known as is to be desired, but the balance of evidence seems to be that the beds are referable to the base of the Permian in the sense in which America basal Permian beds are treated in this paper."

[20] Diller, J. S., *Op. cit.*
[21] Beede, J. W., and Kniker, H. T., Species of the Genus Schwagerina and Their Stratigraphic Significance, Univ. of Tex. Bull. 2433, p. 17, 1924.

In a recent letter, Mr. H. A. Wheeler [22] of Stanford University states that the fauna of the McCloud limestone is lowest Permian and that of the Nosoni formation is probably upper Lower Permian or lower Middle Permian. The abundant fauna of the Nosoni certainly merits further study to determine its exact stratigraphic position.

Triassic

Diller,[23] in mapping the Redding quadrangle, differentiated four members of the Triassic series and briefly described each in the Redding Folio as follows:

Jurassic_____ Modin formation

              Unconformity

Triassic_____	Brock shale	400 feet	Upper
	Hosselkus limestone	250 feet	Triassic
	Pit shale	2000 feet	Middle
	Dekkas andesite	1000 feet	Triassic

              Unconformity

Carboniferous____Nosoni formation

Dekkas andesite.

Dekkas andesite is a thick body of grayish green, green, and deep bottle green flows and interbedded tuffs and breccias. According to Diller,[24] the Permian Nosoni formation is "overlain, most likely with unconformity, by the great mass of andesitic rocks forming Horse, Town, Minnesota, and Salt Creek mountains. The volcanic rocks are nearly or quite alike, and the only guide in drawing the line has been the occurrence of fossils characteristic of the Nosoni formation. No characteristic fossils have been found in this great body of volcanic rocks next above the Nosoni, but in the strata overlying the volcanics characteristic Triassic forms occur, and they are so different as to suggest a decided break somewhere between the Nosoni and Pit formations."

On a later page (p. 8), he further states that "the relation (of the Dekkas) to the Nosoni formation as outlined is equally clear. The outcrops of the Dekkas andesite and Nosoni formations are approximately parallel for over 20 miles, but the Nosoni belt narrows to the south, and on Campbell Creek, as well as near Lilienthal, it is unconformably overlapped by the andesite flows.

Diller also notes that on Minnesota and Salt Creek mountains, tuffaceous slates, interbedded with the lavas and tuffs of the upper Dekkas, contain many microscopic fossils similar to those found in the overlying Pit slates, a statement which scarcely agrees with his comment that the Dekkas contains no characteristic fossils.

The contact between the Dekkas and Nosoni in most places where the writer has examined it between Horse and Town mountains north of the Pit and the Cretaceous overlap to the south is clearly defined.

[22] Wheeler, H. A., Personal communication.
[23] Diller, J. S., *Op. cit.*, p. 6.
[24] Diller, J. S., *Op. cit.*, p. 4.

The Nosoni is composed of clastic sediments which to a large extent are tuffaceous, and a goodly proportion of andesitic flows and dark green, andesitic and basaltic tuffs.   The Nosoni strata are not very resistant to erosion and, where the dips are steep, have had subsequent valleys eroded into them; where the dips are lower, the great mass of resistant Dekkas lavas above has more or less protected them from erosion.   The basal Dekkas consists of massive, grayish green or dark green andesitic flows and firmly lithified pyroclastics of somewhat more salic composition than the Nosoni volcanics.   The Dekkas rocks are very resistant to erosion and form bold ridges and peaks especially north of the Pit River where the formation attains its maximum thickness.   Between the two formations is a distinct erosional discordance, representing the development of a subaerial landscape after the Permian seas had retreated from this region.   No orogenic deformation occurred at this time.   The time interval between the Nosoni and the Dekkas is difficult to estimate. The Nosoni, from studies which so far have been made of its fauna, is not later than lower Middle Permian; the exact age of the basal Dekkas volcanics is not known since no fossils have been discovered in any of the pyroclastics.   The fossils found in the sedimentary interbeds of the upper part of the formation are Middle Triassic in age.

The Dekkas andesite forms a belt 1 to 3 miles wide extending from the Cretaceous overlap in the central part of the Redding quadrangle northward into the Shasta quadrangle.   Splendid exposures are to be found on the north side of the deep Pit River Canyon near the settlement of Heroult.   On Town, Minnesota, and Salt Creek mountains, where the formation reaches a maximum thickness of 1000 feet, excellent sections also are available.   A smaller area of the Dekkas is brought up in an anticlinal fold along Clikapudi Creek south of Copper City in the east central part of the Redding quadrangle.

The Dekkas is composed of flows, tuffs, breccias, and agglomerates in which pyroxene andesite is the dominant petrographic type.   Flows predominate in the basal part while the upper part is comprised chiefly of pyroclastics.   Near the top of the Dekkas, interbeds of tuffaceous shale, closely resembling those of the overlying Pit formation and containing Upper Triassic marine invertebrates, become increasingly abundant and form a transition zone into the dominantly sedimentary Pit. The presence of marine fossils in these interbeds shows that a part of the Dekkas was accumulated below sea level, but there is no evidence that this applies to the whole of the formation.   The abrupt contact of the volcanics with the erosion surface which had been developed in the Nosoni sequence suggests on the other hand that the early Dekkas lavas were erupted over a land surface and the absence of water-laid strata between the flows in the greater part of the formation indicates that subaerial conditions prevailed during most of the epoch.   Towards its close the region occupied by the present southern Klamath mountains gradually sank below sea level and remained submerged during the rest of Triassic time.

Before submergence there had been developed a volcanic plateau surfaced partly by flows and partly by various types of pyroclastics. The area of the plateau is unknown but its original thickness is measured by the maximum thickness of volcanics below the Pit shale for no erosion of consequence appears to have taken place before the Pit submergence and the subsequent burial of the plateau surface by the sediments of that epoch.

The Dekkas lavas are typical pyroxene andesites and basalts which exhibit wide variety of textural and structural features. Extensive chloritization of the ferromagnesian minerals has produced colors ranging from grayish green to deep bottle green. Porphyritic flows with plagioclase and less commonly augite as the phenocrysts, are most common, but fine and even-textured rocks are fairly abundant. Many flows show amygdaloidal and vesicular zones. The pyroclastics are composed chiefly of the same rock types as the flows, though, in addition fragments of acid volcanics and of various types of plutonic rocks are included. Green colors, generally darker than those of the flows, have been developed in the pyroclastics by chloritization of the augite and other ferromagnesian minerals. The microscope shows the rocks to be chiefly highly altered andesites and less quantities of basalt.

Pit formation

As stated above, the Dekkas andesite grades above into the Pit formation by the increase in proportion of tuffaceous slates and shales. The Pit formation, originally called by J. P. Smith [25] the Pitt shale, consists of gray to black shales and slates which are in many cases tuffaceous; lesser quantities of gray, shaly sandstones and sandstone; thinly bedded, impure black cherts; dark gray tuffs; and occasional lava flows. The proportion of pyroclastics, volcanic detritus in the sediments, and lava flows decreases from bottom to top, hence the intense volcanism of the earlier Triassic gradually decreased through the Pit epoch. The Pit was a time of long continued submergence below sea level so that the lava flows and tuffs interbedded with the clastic strata were erupted below sea level and very likely also from vents or fissures close to the coast line. The Pit strata are widely exposed in the Redding quadrangle. Starting as a belt less than two miles wide where they appear at the edge of the Cascade-Columbia lava cap in the Shasta quadrangle, the exposures widen to about ten miles in the highly folded area along the Pit and then decrease to eight or nine miles at the overlap of the superjacent Cretaceous and Cenozoic strata in the central part of Redding quadrangle. The formation has a maximum thickness of more than 2000 feet, though accurate determinations are difficult because of crumpling and faulting.

While the greater part of the Pit formation is non-fossiliferous, poorly preserved invertebrates are present at many horizons, and at a few localities good specimens may be obtained. At the top of the formation are 200 feet of tuff and shale from which Upper Triassic species have been obtained; below at various horizons Middle Triassic types have been collected.

The flows interstratified with the Pit sediments closely resemble those of the underlying Dekkas, that is they are pyroxene andesites in which grayish green or dark green colors have been developed by chloritization of the augite. Petrographically they are identical with the Dekkas types. The tuffs are composed chiefly of crystals of feldspar and of chloritized ferromagnesian minerals. Occasional fragments of slate

[25] Smith, J. P., Op. cit., p. 601.

and of other lavas and plutonic rocks also are present; some slate fragments measure three or four inches across. At several horizons, agglomerates, containing boulders of pyroxene andesite up to six inches across, were found. Gray tuffaceous shales and less common, gray tuffaceous sandstones are abundant throughout the formation; the volcanic constituents are like those of the tuffs described above. These tuffaceous beds at certain localities have been converted to dense, cherty, dark grayish or brownish slates and to dense quartzites. Probably this metamorphism has been caused by the many igneous bodies which were intruded during the late Jurassic deformation which affected this region. These intrusive bodies in some cases show at the surface and metamorphic aureoles follow the contacts; in others their presence must be inferred from the local metamorphism of the Pit strata. Probably under much of the region, igneous masses belonging to this late Jurassic magnatic period are not far below the surface. Metamorphism by deformation alone has not been notable, for in many places little change in the original mineralogy of the rocks has occurred.

Hosselkus limestone

The Hosselkus limestone, lying conformably between the Pit formation and the Brock shale of Upper Triassic age, appears as a series of discontinuous lenses extending from the northern border of the quadrangle near its eastern side southward to the canyon of Cow Creek east of Ingot. The principal exposures are on Brock Mountain, where folding widens the outcrop to nearly two miles. Along the strike this lens extends for about seven miles. The second most important area lies along the canyon of Cow Creek east of Ingot. Other smaller lenses are present north of Brock Mountain. On the north and east slopes of Bear Mountain, 16 miles northeast of Redding and 12 miles southwest of Brock Mountain, there are small exposures of the Hosselkus. The formation according to J. P. Smith [26] is divisible into three lithologic and faunal zones, which are not very sharply differentiated.

Thickness in feet

3. Hard, siliceous, massive, gray limestone which is abundantly fossiliferous (Spiriferina zone) _____ 50
2. Hard, siliceous, prominently jointed, gray limestone containing abundant but poorly preserved fossils (Atracites zone) _____ 100
1. Hard, pure, thinly bedded, dark blue limestone containing abundant well preserved fossils (Trachyceras zone) _____ 50

In the smaller lenses the thickness is less; according to Diller [27] on Cow Creek the thickness is 160 feet, on the east end of Mewittipom Mountain 75 feet, on Bear Mountain about 50 feet. As previously noted the limestone contains abundant fossils, many of which are beautifully preserved, and a very abundant invertebrate fauna has been described by the late J. P. Smith of Stanford University. Cephalopods and corals are the most important types. The fauna shows an Upper Triassic age. Reptilian remains, obtained on both Bear and Brock mountains, have been described by J. C. Merriam.

[26] Smith, J. P., Op. cit., p. 606.
[27] Diller, J. S., Op. cit., p. 5.

Brock shale

The Brock shale conformably overlies the Hosselkus limestone at Bear Mountain and also in the eastern part of the Redding quadrangle. On Brock Mountain, where the principal exposures are located, the lower 300 feet are dark and somewhat calcareous while above are grayish and reddish slates. The shales contain a small Upper Triassic fauna.

The principal exposures form a narrow belt extending from the northern margin of the Redding quadrangle southward for about 15 miles. A small area shows in the hilly country north of the canyon of Little Cow Creek and east of Ingot.

In the closing epoch of Triassic time, occasional eruptions of andesitic magma again occurred, for in the lower part of the Brock formation which overlies the Hosselkus are thin beds of andesitic tuffs and tuffaceous shales. In exposures of the upper Brock strata seen by the writer, volcanic detritus does not appear and Diller does not record its presence in the more extensive areas which he studied. So far as the record shows, volcanic activity was not very intense and died out early in the Brock epoch. The Brock tuffs and tuffaceous shales are similar to those of the Pit formation, that is they are andesitic in composition.

On Bear Mountain Diller suggests the presence of Brock shale, but did not identify it. Between the Hosselkus limestone and the basal Modim agglomerate, there are about 40 feet of thinly bedded, dark gray shales and tuffaceous shales closely resembling the Brock shale farther to the east which certainly must be assigned to the formation. Thin beds of andesitic tuff also are present.

A few species of pelecypods and cephalopods are fairly abundant in most of the Baird exposures.

Jurassic

A. Pre-Nevadan

In the Redding quadrangle, Diller [28] divided the very thick and areally extensive Jurassic sequence into three formations, the Modin below, the Bagley andesite, and the Potem above.

1. Modin formation

The lowest member of the Jurassic in the southern Klamath was named the Modin formation by Diller [29] from exposures near the mouth of Modin Creek in the northeastern part of the Redding quadrangle. The formation comprises "tuffaceous beds, overlain by a greater mass of compact, fine gray shaly sandstones and shales, with a few small lentils of limestone." At the base is a thick volcanic agglomerate consisting chiefly of andesite debris but containing also abundant fragments of the underlying Triassic formations notably the Hosselkus fossiliferous limestone. The volcanic boulders rarely exceed one foot in diameter. On Bear Mountain about sixteen miles northeast of Redding this agglomerate zone is about 400 feet thick. The sandstones are fine-grained, gray in color, commonly thinly bedded, and have associated with them shaly layers. Thin limestone lentils, 10 to 20 feet and 12 to 18 inches thick are interbedded with the sandstones. Near Bagley

---

[28] Diller, J. S., op. cit., pp. 5–6.
[29] Diller, J. S., op. cit., p. 5.

Mountain, in the northeastern part of the Redding quadrangle, are larger limestone bodies, one of which is a mile long and 200 feet thick. The principal exposures form a belt two to four miles wide extending from the north side of Cow Creek in the eastern part of the Redding quadrangle to its northern border about seventeen miles distant. The belt continues for a few miles still farther north into the Shasta quadrangle where it passes beneath the Cascade lavas; on the south, it continues into the Lassen quadrangle and again disappears beneath the Cascade lavas. A small, isolated area of the basal volcanics is exposed on the top of Bear Mountain in the central part of the Redding quadrangle south of the Pit River. Diller estimated the thickness of the Modin at about 3000 feet, but does not indicate that a detailed section was measured. On Bear Mountain the volcanics have a thickness of about 400 feet.

The break between the Modin and the underlying Brock in general is slight for the narrow belt of Brock between the Hosselkus and the Modin runs for miles through the northeastern part of the quadrangle with very little change in width. None the less, a distinct erosional and in places a slight angular discordance exists. On Bear Mountain as noted previously, about forty feet of shales and thin tuffaceous beds which certainly belong to the basal Brock since pyroclastics and pyroclastics sediments have not been reported in the upper part of the formation. Between the Modin and the Brock on Bear Mountain is a slight angular and erosional unconformity, hence a moderate deformation of the crust took place in that section. Because of the limited area of the outcrop and its isolated position, evidence favoring either view can not be obtained.

Regarding the age of the Modin, Diller [30] quotes Stanton as stating that fossils from the basal " 'tuff' are of Upper Triassic (Hosselkus) age, but the tuff in which the limestone is found is more probably of Jurassic age. Between the Pseudomonotis horizon (Brock shale) and the beds yielding a well characterized Jurassic fauna (Potem horizon) comparable with that found at Taylorsville, there is a broad belt in which a great thickness of rocks is represented and from which fossils were obtained at many localities. In the field these were considered Jurassic, and I still think that most if not all of them are of that period, but the paleontologic evidence is not so complete as is desirable. The fossils at most localities belong to types that would not aid in discriminating Jurassic from Triassic." Crickmay [31] implies that the Modin should be classed as Liassic in age, since he states that the Bagley volcanics overlie Liassic deposits, though he does not mention the Modin directly.

On Bear Mountain the basal volcanics alone are represented; if the much thicker upper sedimentary part ever were present it has been eroded away. At the base is a coarse agglomerate which consists chiefly of fragments of dark red and black andesites and basalts and great angular fragments of slate and limestone evidently derived from the underlying Triassic embedded in a matrix of fine rock fragments and shattered crystals of plagioclase, hornblende and quartz. The rock fragments range from one-eighth of an inch to two feet in diam-

---

[30] Diller, J. S., *op. cit.*, p. 5.
[31] Crickmay, D. H., Jurassic History of North America, Proc. Amer. Philos. Soc., vol. 70, p. 28, 1931.

eter. Higher in the section, the tuffs are finer grained, better stratified, and few conspicuous rock fragments are visible. A single thin bed of gray shale is present near the top of the section. The thickness of these volcanics is about 400 feet.

According to Diller, splendid sections of the entire Modin formation may be found at many places along and north of the Pit River. Between Flat and Potem creeks in the Pit canyon, the basal agglomerate overlies the Brock shales and contains fragments of fossiliferous Hosselkus limestone as on Bear Mountain. Several other agglomerates are present at higher horizons. The upper sedimentary section is well exposed along the Pit.

2. *Bagley andesite*

Under the name Bagley andesite, Diller [32] describes two areas of volcanic rock lying between the Modin and Potem formation, "one from which the rock takes its name, near the northeast corner of the Redding quadrangle, and the other along the Pit River near the eastern border." More than three-quarters of the first area is composed of "tuff" which Diller describes as "sometimes coarse, almost agglomeratic, but generally fine and stratified, with occasional traces of marine fossils."

Lavas are most abundant near the summit of Bagley Mountain. Diller classes the rocks as andesites, but in his descriptions says nothing about the composition of the feldspars. In the deep canyon of the Pit near the mouth of Potem Creek a magnificent section of the second Bagley area is exposed. Here flows and tuffs are present in more nearly equal proportions, and contemporaneous dikes and sills cut the extrusive rocks. By chloritization of the augite and other ferro-magnesian minerals, the rocks of the Bagley have been converted to greenstones.

Diller states that "both areas lie practically on the border between the Modin and Potem formations, but do not necessarily indicate an unconformity. These areas represent centers of greater accumulation of volcanic material near points of eruption near the beginning of the Potem epoch. Between these two points the contemporaneous sediments contain some detritus from both centers, but apparently the greater portion comes from a different source. For this reason the intermediate sediments were included in the Potem."

A brief description of the petrographic characters of the Bagley rocks is given by Diller [33].

3. *Potem formation*

Overlying the Modin formation or the intervening Bagley andesite where the latter is present is the Potem, a formation also named by Diller [34] from the type locality on Potem Creek in the northeastern part of the Redding quadrangle. "The Potem formation is composed of sandstone, shales, and tuffs. The thin-bedded sandstones and gray, sometimes slaty shales predominate in its lower part and make up the greater part of the formation. They are more or less calcareous and contain a few lentils of limestone. Tuffaceous conglomerates occur

[32] Diller, J. S., *op. cit.,* p. 8.
[33] Diller, J. S., *op. cit.,* p. 8.
[34] Diller, J. S., *op. cit.,* p. 5.

sparingly in the lower half of the formation, but in the upper part they are most abundant—in fact, nearly all of the sediments in this part are of igneous material. Some of this may have been furnished by contemporaneous volcanic activity, but most of it was derived by the ordinary processes of erosion from a wide expanse of volcanic rocks." As a matter of fact, limestones, shales, and sandstones are fairly abundant in the upper part of the formation, and many of the limy beds contain rather abundant fossils. Whether all of these sediments contain volcanic detritus certainly has not been determined; many of the beds evidently contain very little. The tuffaceous conglomerates of which Diller speaks are chiefly coarse and fine tuffs, though some agglomerates are present.

The Potem formation is exposed over a considerable area in the northeastern part of the Redding quadrangle and extends into the Modoc Lava Beds quadrangle as far as the Kosk Creek drainage, beyond which it is buried by the great pile of Cascade lavas; magnificent exposures are to be found in the rugged hills of this region so that the stratigraphy can probably be worked out in very considerable detail. To the south the Potem extends into the Lassen Peak quadrangle where again it is buried by the Cascade lavas. Diller estimates the thickness of the formation at about 2000 feet, but gives no details of measured sections. Regarding the age, Diller [35] quotes Stanton as stating that the Potem is "characterized by a well-marked Jurassic fauna including several species that occur in the Hardgrave sandstone near Taylorsville. Among the forms may be mentioned Rhynconella, Pecten acutiformis Meek, Pecten (Entolima), Pinna expansa Hyatt, Modiola, Gervillia, Lima, Trigonia (several species), Goniamya, Pholadomya, and ammonites of the Coroniceras types, etc. It is probable that horizons somewhat higher than those of the Hardgrave sandstone are included in the formation, but some of the types that seem later, as indicated by their occurrence in the Taylorsville region, are intimately associated with the Hardgrave species. This fauna is very distinct from all those that precede it in this region." Hyatt classed the Hardgrave sandstone as Upper Liassic and Crickmay [36] more definitely dates it as Ludwigian. Crickmay writes that the lower part of the Potem consists of sandstone and shale and contains presumably Middle Jurassic fossils," and the upper part of "tuffs and agglomerates which mark the later Middle Jurassic volcanism."

The igneous rocks of the Potem are chiefly coarse and fine andesitic tuffs; agglomerates are present in a few places and there are occasional andesitic and basaltic flows interbedded with the clastics. Sediments (limestones, sandstones, and shales) containing various proportions of volcanic detritus are abundant especially in the upper part of the formation. These various pyroclastics and flows prove the recurrence of volcanism in the Middle Jurassic after the lapse represented by the dominantly sedimentary phase of the lower Potem. The region apparently stood below sea level during Potem time so that the pyroclastics were either erupted under water or were blown into the oceans by eruptions on adjacent land areas. Volcanics erupted on the lands probably supplied most of the debris present in the sedimentary beds

[35] Diller, J., Geology of the Taylorsville Region, California, U. S. Geol. Surv., Bull. 340, p. 40, 1908.
[36] Crickmay, C. H., Jurassic History of North America, Proc. Amer. Philos. Soc., vol. 70, p. 29, 1931.

of the lower and upper Potem. The Middle Jurassic thus was a time of intense volcanic activity in the Klamath and also in the Sierra Nevada region where pyroclastics and flows were accumulated over large areas in considerable thicknesses.

Late Jurassic (Nevadan) intrusives

During the later part of the Jurassic, the earth's crust in the Klamath-Siskiyou region was strongly folded and faulted; mountain ranges of some magnitude were developed. As this deformation progressed large volumes of granitoid magma were intruded into the crust and solidified chiefly as various types of diorites. Later parts of this intrusive complex were exposed by erosion as dikes, sills, chonoliths, stocks, bosses, and batholiths. Those included within the boundaries of the mapped area are briefly described.

1. *Plutonic bodies*

The largest bodies include a small "batholith," numerous stocks and bosses, and a great, irregular plutonic dike all of which apparently represent the higher domes and ridges of a granitoid complex present under much of this region. Farther north and west in the northern Klamath and in the Siskiyou ranges are other similar granitoid bodies probably intruded at the same time and belonging to the same complex.

a. *Trinity Alps stock-boss cluster*

North of Weaverville, the county seat and chief town of Trinity County, in the high, rugged mountains which are locally known as the Trinity Alps, is a cluster of stocks and bosses of quartz diorite. The stocks form high peaks which are specially conspicuous in the landscape because of the sharp contrast between the light gray color of their rocks and the generally dark shades of the country rocks which are mainly serpentine and hornblende and chlorite schist.

In the rugged walls of the many deep canyons in this section are magnificent exposures of the contacts of the stocks with the country rock. Where their downward extensions are visible, the walls dip away from the rounded summits of the stocks at angles generally exceeding fifty degrees. Many apophyses extend outward from the main bodies, and bays and re-entrants in the walls are abundant so that in detail most of the contacts are very irregular. Where the magma invaded the Salmon schist, offshoots in great numbers have intruded the schist either along the planes of schistosity or at various angles thereto, and thus have developed contact zones of exceedingly complicated structure. Lit-par-lit gneisses have been developed at many places.

The rocks of the various stocks and bosses are pale gray quartz diorites composed of plagioclase, quartz, hornblende, and biotite with accessory magnetite visible under a lens. Chlorite and the weathering products of plagioclase also are visible. The texture is generally coarse and equigranular, but fine grained areas and contact zones are not uncommon. In certain areas, the large size of the feldspars and less commonly quartz and hornblende individuals gives the rock an inequigranular texture. Along some of the contacts a considerable variety of hybrid rocks have been produced by reaction between the magma and the country rock. Autoliths are fairly common in the larger bodies. The microscope shows that andesine composes considerably more than half of the rock, with quartz, green hornblende, brown biotite, and accessory magnetite, apatite, zircon, and titanite making up the residue. Chlorite, epidote, zoisite, calcite, kaolin, and sericite are secondary products.

Aplite and pegmatite dikes are rather abundant in these bodies and extend into the adjacent country rock.

b. *Shasta Bally batholith*

The largest subjacent body in the region lies principally within the southern half of the Weaverville quadrangle, but extends a short distance into the adjacent Red Bluff quadrangle where it passes beneath the superjacent Cretaceous deposits. The exposed part of the batholith is 30 miles long and 10 miles in maximum width; at its northern end it narrows to an irregular dike-like body. In terms of Daly's [37] definition of stocks and batholiths, this intrusive belongs to the second group since its area considerably exceeds the 100 square kilometers which Daly sets as the upper limit for the area of a stock.

In the Shasta Bally batholith there are two rock types which grade into each other—hornblende biotite quartz diorite and hornblende quartz diorite containing a very small percentage of biotite. Some field observers have attempted to establish the existence of two bodies, but a careful examination of the exposed area shows that this is not possible. Along certain contacts, notably with the Mississippian Bragdon slates and the Copley meta-andesite, the rock is medium to coarsely granular biotite-hornblende quartz diorite; along part of the contact with a serpentine intrusive, this phase is fine-grained and in places shows especially large feldspars. The main mass is lighter colored and the percentage of biotite is small; the two types grade insensibly into each other. The biotite-rich phase appears to have been developed along contacts by the assimilation either of Mississippian slates or of basic igneous rocks.

The typical hornblendic phase is a light or medium gray rock of coarse and generally even texture. In the hand specimen plagioclase, quartz, hornblende, biotite, and magnetite are visible. The microscope shows that the feldspar, ranging from basic to acid andesine, composes more than 60 per cent of the rock. The residue consists of quartz, green hornblende, green biotite, and accessory magnetite, titanite, and apatite.

c. *Mule Mountain stock*

This body lies along the common border of the Redding and Weaverville quadrangles, and is separated from the Shasta Bally batholith by a narrow belt of Copley meta-andesite and Balaklalla granite porphyry. The stock is eight miles in length and four and three-quarters miles in greatest width; its boundaries are exceedingly irregular. The rock is a highly quartzose diorite which locally is quite deficient in ferromagnesian minerals. The texture is generally equigranular, but, in certain quartz-rich phases, many of the quartz grains are considerably larger than the average grain size of the rock. The principal

---

[37] Daly, R. A., Igneous rocks and their origin. New York, p. 90, 1913.

development of this phase is along the indefinite contact with a granite porphyry chonolith which it has intruded.

The quartz diorite is so deeply weathered that fresh specimens are difficult to secure.

The principal rock type is composed of albite or albite-oligoclase, quartz, green hornblende, small amounts of green biotite. A small amount of orthoclase is present. Magnetite, titanite, and scattered crystals of apatite are the chief accessories. Kaolin, sericite, and calcite are alteration products of the feldspars; epidote, chlorite, zoisite, and calcite of the hornblende and biotite.

A few deeply weathered autoliths and scattered pegmatite and aplite dikes are present.

d. *Pit River stock*

The Pit River stock is a small body of quartz diorite about three miles long and one and one-half miles wide which is well exposed in the canyon of the Pit River near the Pacific Highway bridge.

The typical rock is a medium to coarse grained light gray rock with certain phases showing specially large quartz grains which give it a pegmatoid structure. Plagioclase, often pale pink in color, and chloritized ferromagnesians, epidote, magnetites, and chlorite, are the other megascopic minerals.

Aplite dikes are common and the rocks are almost white, sugary textured, and contain little or no visible ferromagnesian minerals.

e. *Redding quartz augite diorite plutonic dike*

One of the most interesting plutonic intrusives in the southern Klamath Mountains is a huge, irregular, dike-like body of quartz augite diorite which extends far more than 20 miles southward from the northern border of the Redding quadrangle along or close to the McCloud River. How far it projects into the Shasta quadrangle is not known, but, judging from the narrowness of the dike at the boundary of two sheets, the exposed distance can not be great. The southern end continues for about two miles south of the limit set by Diller, so the total length of the dike must exceed 25 miles. The maximum width is two and four-tenths miles and the average width about one mile. Throughout most of its course the magma rose along the contact between the Mississippian Baird formation and the Permian McCloud limestone, but, during the process of intrusion, active stoping took place so that the contacts are distinctly irregular. The McCloud limestone in particular was greatly shattered and, along the course of the dike, are many xenoliths and roof pendants of this rock, and also of the Baird and Nosoni rocks.

The principal contacts as stated above are with the Mississippian Baird formation and the Permian McCloud limestone, but in a few places the dike transgresses the Permian Nosoni formation and the early Triassic Dekkas formation. At two localities, there are fairly long stretches of contact with the Dekkas andesite, the youngest formation cut by the main dike. One of these extends from the North Gray Rock for over two miles to the north side of the Pit River canyon; the second extends from Dekkas Creek also for two miles to the south side of Campbell Creek. Dike-like and irregular apophyses can be traced from both sides of the main body and many isolated intrusives of the same rock type are present close or at some distance from the main dike. On the west the dike and its offshoots cut principally Mississippian formations; on the east side the youngest formations which the apophyses intrude are the Triassic Pit and Jurassic Modin formation, but of course the relations of these bodies to the main one can not be proved. The dike certainly is post-Dekkas and probably belongs to the Nevadan intrusives though it was intruded before the Pit River stock which has been previously described.

The rock of the Redding dike exhibits a considerable range of color, texture, and mineral composition. The most common type is a medium- and generally even-grained, dark-green rock in which greenish plagioclase, chloritized augite, quartz, and magnetite are visible in the hand specimen. In certain areas, especially large plagioclase individuals give to the rock an inequigranular texture. Less common is a somewhat coarser grained, medium grained rock in which the plagioclase is white. Along portions of the contacts and in apophyses, the rock is fine grained, often inequigranular, and possesses colors ranging from grayish green to dark bottle green. This last phase very closely resembles meta-andesites of the Nosoni and Dekkas formations which are cut by the Redding dike and its apophyses, hence care must be used in the field in properly determining the relations.

Examination of thin sections of typical specimens show andesine, chloritized diopsidic augite, and interstitial quartz as the chief constituents. As accessories, magnetite, apatite, and zircon were present in all sections examined, while green hornblende, almost completely chloritized biotite, and ilmenite are present in certain sections.

Alteration products of the ferromagnesian minerals are chlorite, epidote, zoisite, and calcite; of plagioclase, calcite, zoisite, and serpentine. Most sections examined show extensive alternation.

A fine-grained phase from a five-foot dike cutting the McCloud limestone on the north side of Potter Creek canyon is composed of andesine, quartz, chloritized augite, magnetite, and additional secondary minerals, zoisite, epidote, and calcite. The texture is porphyritic with plagioclase the phenocrystic mineral. In many of the small offshoots, augite also appears as phenocrysts.

2. Hypabyssal intrusives

a. *Chonoliths of soda granite porphyry* (alaskite porphyry)

The largest hypabyssal bodies of probable Nevadan age are two chonoliths of soda granite porphyry, one of which, the Balaklala chonolith, forms the main mass of the ridge extending from Iron Mountain for some distance north of Behemotash Mountain west of the Sacramento River, and the other, the Bully Hill chonolith, is exposed along the lower slopes of Horse and Town mountains near Bully Hill north of the Pit River. A large area of the second body is exposed in the hilly and mountainous country in the eastern part of the Redding quadrangle south of the Pit River, and a small isolated mass extends for about four miles north of the Pit from near the mouth of Clikapudi Creek. A still smaller section extends from the head of Clover Creek into the Lassen quadrangle and others of more limited extent are present at various places in the eastern part of the Redding quadrangle.

Diller [38] originally described these bodies as two series of rhyolite flows and ejecta. The Balaklala mass he believed to underlie the middle Devonian Kennett formation and the Bully Hill complex to have been erupted prior to the deposition of the late Triassic Hosselkus limestone. Diller noted certain intrusive contacts but held that the masses were chiefly extrusive.

L. C. Graton,[39] during an investigation of the copper deposits in Shasta County, established the intrusive character of both bodies and this Ferguson [40] accepted in his report on the gold lodes of the Weaverville quadrangle. Graton's report describes the petrography of the rock which he calls alaskite porphyries and gives an excellent summary of their intrusive character. The writer has examined practically all of the contacts of both bodies and in no place can Graton's findings be disputed. Magnificent exposures of the contacts of the Balaklala body are visible in many places along the steep slopes of the canyons which have been carved into the eastern side of Behemotash Mountain where the main mass of the intrusive is exposed and, on the western side of this ridge, equally fine outcrops may be found. Near Winthrop, especially along the railroad cut leading to the Rising Star Mines, the intrusive character of the Bully Hill body is clearly shown, and at many other localities also similar relations are easily identified. Graton (p. 87) suggests that the intrusives may be laccolithic since "in the western districts the Kennett and Bragdon formations have been arched up by the alaskite porphyry (Balaklala body), and in the eastern districts the Pit formation has been deformed by it," though he does not consider it certain "that true laccolithic structure is present." In the writer's opinion, both masses should be classed as chonoliths rather than as laccoliths since (1) they considerably exceed the size set by Gilbert as the maximum for true laccoliths—the dimensions of the Balaklala body are about 6 by 13 miles, of the Horse and Town mountain section of the Bully Hill body 1 by 4 miles, and of the larger section in the eastern part of the quadrangle about 5 by 9 miles; (2) they are irregular in shape; (3) while their contacts are concordant in part, the principal stretches are distinctly discordant; and (4) considerable lengths of contact of the Balaklala body are against essentially massive igneous rock.

The rock composing the two chonoliths normally possesses a porphyritic texture with phenocrysts of quartz and feldspar or of feldspar alone in a predominant, fine-textured groundmass; along certain contacts and also within the bodies are areas of dense, fine-grained rock. The Bully Hill rock generally has a dark gray groundmass while that from the Balaklala body ranges from dark gray to pale green or greenish-gray. The porphyries vary considerably in appearance from place to place. Where the Balaklala porphyry adjoins the Mule Mountain quartz diorite stock, the texture becomes exceedingly coarse and a gradual transition between the rock types of the two bodies exists so that a definite boundary between the two can not be located. On the map, the approximate boundary is indicated by the dotted line but the pattern does not show the transitional nature of this contact. Auto-

[38] Diller, J. S., The Redding Folio No. 138, pp. 7, 8, 1906.
[39] Graton, L. C., The Occurrence of Copper in Shasta County, California. U. S. Geol. Surv., Bull. 430, pp. 82-83, 1910.
[40] Ferguson, H. S., Gold Lodes of the Weaverville quadrangle. U. S. Geol. Surv., Bull. 540, p. 30, 1914.

brecciated zones are present at a number of places and probably are the phases which Diller interpreted as pyroclastics. Some of the chief variations in these chonolith rocks have been produced by shearing, which is a conspicuous feature in both bodies though probably is more pronounced in the Balaklala chonolith. Along the shear zones the porphyry takes on a schistose structure and in many places has gone over to a highly foliated quartz sericite schist.

At the surface in most places, weathering has bleached the rocks to various shades of pink, red, yellow, brown, gray and green. The porphyries weather easily and altered rock extends scores of feet below the surface. In general the porphyries weather much more easily than any of the rocks which they intrude. This contrast is well shown between the Balaklala rock and the Copley meta-andesite which it intrudes in exploratory tunnels driven by U. S. Army Engineers during investigations connected with the possible location of a dam across the Sacramento River a few miles south of Kennett. The oxidized zone in the Copley meta-andesite rarely extends more than five or ten feet below the surface while in a tunnel which went into the Balaklala porphyry at least 150 feet below the surface, that rock was completely altered. How far below the tunnel floor completely or partially altered rock extends is not known. Deep alteration of the porphyries by meteoric waters is shown in the many mines in both the Balaklala and Bully Hill chonoliths.

The microscope shows phenocrysts of quartz, albite, and rarely oligoclase-albite are generally of small size though in a few places they may attain a diameter of one-quarter of an inch or even more; in such places the groundmass also commonly increases in coarseness. The fine-grained groundmass consists of a microgranular mixture of quartz and feldspar. Fine shreddy grains of magnetite and small particles of chlorite and of epidote that were probably derived from original flakes of biotite constitute the chief accessory constituents, but are nowhere present in important amounts. Apatite, titanite, and zircon exist sparingly as small crystals.

One of the specially striking lines of evidence proving that the porphyries are intrusive is the contact metamorphism of both the Devonian Kennett and Triassic Pit shales. At many localities these shales have been highly indurated and converted into dense hornfels, or they have been rendered porous and slaggy by the intense heat of the invading bodies. On the steep, western slopes of the Sacramento Canyon, contact effects of the first type have been developed in the Kennett shales and north of Winthrop at Bully Hill is a splendid exposure of slaggy Pit shale which has been changed from the normal dark gray to a dark brick red color. Studies of mineralogical and structural changes in the shales so far have not been made.

b. *Dikes, sheets, and other small bodies*

Great numbers of dikes, sills, and small irregular bodies of intrusive rock are present in the vicinity of the larger plutonic and hypabyssal bodies and undoubtedly represent offshoots from them. Isolated intrusions at some distance from any major bodies are composed of similar rocks and doubtless were injected into the crust at the same time. The dikes and sills occur alone or in groups or swarms. Within the stocks and bosses are many pegmatite and aplite dikes, which have already been described; the pegmatites in many places pass over into veins composed almost wholly of quartz. In color, texture, and mineral composition, the satellitic bodies exhibit a wide range. The principal petrographic types are quartz diorite, granodiorite, quartz augite diorite, soda granite porphyry, dacite porphyry, diorite porphyry, andesite, basalt, and lamprophyre. For brief petrographic descriptions of some of the types, the reader is referred to published reports by Diller,[41] MacDonald,[42] and Ferguson.[43] Further data will be presented in a later report by the writer on the Weaverville quadrangle.

Contact metamorphism about the larger Nevadan intrusives

About the various Nevadan plutonic intrusives and the two granite porphyry chonoliths, metamorphic zones and aureoles have been developed and in many places are magnificently exposed; particularly fine contacts are to be observed between the Nevadan Redding dike and the McCloud limestone and between Nevadan granodiorite and quartz diorite stocks and various rocks in the deeply glaciated sections of the interior Klamath. Because of the wide variety of rock types invaded by the various magmas, many different suites of metamorphic rocks and minerals are present, and unusual opportunity is offered for the field study of manifold contact phenomena. Profound metamorphism about igneous contacts has occurred in the limestones, especially the Permian McCloud, the Mississippian Bragdon shales, the Copley meta-andesite, the serpentines, and in a few places the Salmon and Abrams schists. The best developed and widest aureole surrounds the Shasta Bully batholith, the largest Nevadan plutonic body in the area described, but even more striking metamorphism has been produced in the McCloud limestone by the Redding quartz augite diorite dike. Moderately wide contact metamorphic zones, developed at a higher level in the earth's crust and therefore of different type, are present along the contacts between the Balaklala and Bully Hill granite porphyry chonoliths and the Devonian and Triassic shales respectively. About most of the stocks more or less conspicuous zones also are present.

THE SUPERJACENT FORMATIONS

All the formations previously described were strongly folded in the later part of Jurassic time and subsequently were deeply eroded. Upon this deformed and dissected crust lie the supercuit strata of Cretaceous and Cenozoic ages.

Post Nevadan Upper Jurrassic

Knoxville formation

The strata immediately overlying the subjacent complex in the area mapped have been divided into three important geologic units, the Knoxville at the base, the Horsetown, and the Chico. By some authorities (Becker, C. A. White, Whitney, J. P. Smith, F. M. Anderson, B. L. Clark, and others), the Knoxville has been considered partly

[41] Diller, J. S., op. cit., pp. 7-9.
[42] McDonald, D. F., Notes on the Gold Lodes of the Carrville District, of Trinity County, California, U. S. Geol. Surv., Bull. 530, pp. 11-13, 1912.
[43] Ferguson, H. S., op. cit., pp. 30-33.

or wholly Upper Jurassic. Many years ago, Diller and Stanton [43a] studied a section along Elder Creek in Tehama County south of the area under discussion, and divided it as follows:

Upper Cretaceous—Chico formation_____ 4,000 feet

Lower Cretaceous { Horsetown formation_____ 6,000 feet
{ Knoxville formation _____ 20,000 feet

Others have followed the lead of Diller and Stanton in classifying the entire section as Upper Cretaceous, and this usage has been accepted by the United States Geological Survey. [43b]

F. M. Anderson, an early student of these deposits, has recently reviewed their field relations and their paleontology, and has concluded from stratigraphic and faunal evidence that the thickness of the Knoxville must be considerably reduced, that the formation is Upper Jurassic, and that an important unconformity separates it from the overlying Cretaceous. The Cretaceous, including many thousands of feet of beds formerly classed as Knoxville, he divides into the Lower Cretaceous Shasta series, composed of the Paskenta and Horsetown formations, and the Upper Cretaceous Chico formation. The following data from a section studied near Paskenta in Tehama County have been kindly furnished by Dr. Anderson; [43c] his detailed findings will shortly appear in printed form.

Upper Cretaceous—Chico formation_____ 4,000 feet

Lower Cretaceous

Shasta series { Horsetown formation _____ 7,660 feet
{ Paskenta formation _____ 4,880 feet

Unconformity

Upper Jurassic—Knoxville formation_____14,280 feet

In the area covered by the present report, the boundaries between the Knoxville and Paskenta formations and between the Paskenta and Horsetown formations have not yet been determined. According to Anderson, the Knoxville north of Paskenta is progressively overlapped by the Horsetown so that its exposed thickness is rapidly decreased. Whether true Knoxville lies in contact with the Klamath subjacent complex in the northwestern part of the Sacramento Valley is not yet known, but there is a considerable thickness of strata closely resembling those of the Knoxville farther south which the writer tentatively correlates with that formation. No fossils have been found to aid in their identification. Both horizons of the Shasta series, the Paskenta and the Horsetown, have been recognized, and each is represented by many thousands of feet of strata.

Below the lowermost Shasta strata in the northwestern part of the Red Bluff quadrangle, mostly south of the Redding-Platina road, is an undetermined thickness of conglomerate, sandstone, and shales which resemble the Knoxville of the type section farther to the south. No fossils have been found in these rocks, hence their correlation with the

[43a] Diller, J. S., and Stanton, T. W., Shasta Chico Series, Bull. Geol. Soc. Amer., vol. 5, pp. 438–422, 1894.
[43b] Wilmarth, M. Grace, Names and Definitions of the Geologic Units of California, U. S. Geol. Surv., Bull. 826, p. 39, 1931.
[43c] Anderson, F. M., Personal communication.

Knoxville except on similarity of lithology has not been established, but the existence of a considerable thickness of strata below the lowest Shasta horizon yet discovered is strongly indicative of Knoxville age. At the base of this sequence are local conglomerate lenses composed of boulders and pebbles derived from adjacent bedrock, beds of lignite and lignitic shale which lie close to the base and below some of the thickest of conglomerates. Above are gray shales and some brown sandstones. These strata are not found in either the Red Bluff or the Weaverville quadrangles. They lie with strong unconformity on the eroded edges of the subjacent series.

Cretaceous

Shasta series. Paskenta and Horsetown formations.

In the Weaverville quadrangle, the Paskenta division of the Lower Cretaceous Shasta series is exposed along Brown and Redding creeks about 5 miles southeast of Douglas City and also in the southwest corner on the western slope of Hayfork Mountain. In the Brown and Redding Creek area, about 1,100 feet of strata overlie the Abrams schist and dip sharply to the south. The basal third is composed chiefly of medium- to fine-grained, buff, firmly cemented, arkosic sandstone with occasional layers of conglomerate and gray shale. The upper two-thirds consist chiefly of dark gray shale with thin beds of sandstone and conglomerate. Fossils, which are rather abundant in the sandstones, are considered by F. M. Anderson[43d] to belong to the middle portion of the Paskenta.

In the Red Bluff quadrangle, a great thickness of Shasta strata overlies the Subjacent complex, and is separated from it by a pronounced erosional and angular unconformity. The stratigraphy of the Shasta series in this region is only partially known; neither the base nor the top of the Paskenta have been determined. The separation between the Horsetown and the Upper Cretaceous Chico formation has been established. On the geological map accompanying this report, the Lower Cretaceous in the Red Bluff and Weaverville quadrangles is designated as Shasta.

The Shasta series generally consists of a basal conglomerate composed of pebbles and boulders of the adjacent bedrock, coarse arkose, brownish and greenish, commonly arkosic sandstones varying in texture from coarse to fine, and dark gray shales; the arkose and sandstones for the most part are present in the lower part of the formation above the basal conglomerate, but are also found throughout the formation. In the principal area in the Weaverville quadrangle on Brown and Redding creeks, brown sandstone unconformably overlies the Abrams schist and is succeeded by dark gray shales which comprise the upper two-thirds. The thickness of this exposure is about 1000 feet. In the Red Bluff quadrangle the base of the Shasta series has not been determined, though recently discovered fossils by F. M. Anderson and R. Dana Russell show that the base is several thousand feet below that set by Diller and Stanton. Thus the total thickness of the Horsetown in the Red Bluff quadrangle is not known, but probably is more than 6000 feet.

[43d] Anderson, F. M., Personal communication.

From a short distance west of Igo in the Weaverville quadrangle to Jerusalem Creek in the Red Bluff quadrangle (a distance of eight miles) the Horsetown division overlies the Shasta Bally quartz diorite batholith.   At the base, the Horsetown is composed of medium gray, massive coarse-grained, fresh arkose composed of plagioclase, quartz, biotite, and hornblende in about the same proportions as in the quartz diorite.   Poorly preserved fossils are present at a few horizons in this arkose.   Locally conglomerate, composed chiefly of quartz diorite pebbles and boulders with lesser numbers from the other subjacent formations, is present at the base or at higher horizons interbedded with the arkose.   Above the main arkoses and conglomerates are greenish and brownish arkosic sandstones commonly containing much biotite, and these in turn are succeeded by gray shales with occasional sandstones and conglomerates.   Some zones are highly fossiliferous.   This section is Upper Horsetown.

At Byron Gulch, where Byron Stream joins the North Fork of Cottonwood Creek a lower section of the formation is exposed.   At the base is a conglomerate ranging from 250 to 300 feet thick, which is poorly sorted and is composed of pebbles and boulders some of which are more than two feet in diameter.   The pebbles and boulders are composed of rocks from various subjacent formations such as the Shasta Bally quartz diorite, Bragdon conglomerates and slate, Copley meta-andesite, and various other igneous rocks.   Above are massive sandstones with occasional shaly interbeds, interbedded sandstones and shales, sandy shales, and finally dark gray, nodular, carbonaceous shales.   Intraformational conglomerates, fine- and coarse-grained sandstones containing shale fragments are present at various horizons in the dominantly shaly upper part of this section.   Fossiliferous zones are numerous.

Chico formation

The Chico formation is widely exposed in the Redding quadrangle along the southern Klamath foothills which it partially buries; it does not show in the Weaverville area.   It overlies the subjacent series with strong angular and erosional unconformity, and dips at angles of less than twenty degrees to the south or southwest.   At the base there is generally a rather coarse to very coarse conglomerate composed principally of rounded and subangular boulders of the subjacent formations immediately underneath; where the Chico is in contact with the Triassic, the conglomerate consists of generally sharply angular fragments of the slates.   Thus the character of the boulder beds varies along the strike depending on the nature of the bedrock.   The size of the boulders in the basal conglomerate also vary from place to place; most do not exceed eight inches in maximum diameter, though occasional boulders are two feet or more across.   Above are coarse, arkosic, firmly cemented sandstones, general yellowish or brownish in color, and contain considerable volumes of dark mineral and rock fragments; these are succeeded by finer-textured sandstone of same composition. Shales and sandy shales, some of which are carbonaceous or even lignitic, are interstratified with sandstone.   Most of the sandstones are massive and thick bedded; but some exhibit moderately thin layering.   Clay ironstone concretions of dark brown color are commonly present; these either stand out from the surface or have been removed leaving pits. The sandstones become more or less abundant and shaly in the upper portion of the formation which, in the central part of the quadrangle, is composed of gray shales generally containing clay ironstone concretions. On the eastern side of the Sacramento Valley in the Redding quadrangle, at the top of the Chico are brown, arkosic sandstones and conglomerates in which some boulders reach four feet in diameter.

The principal exposures of the Chico are in the eastern part of the Redding quadrangle along South Cow Creek, Old Cow Creek, Basin Hollow Creek, Clover Creek, Oak Run and Swede Creek. A small area is located near Oak Flat and one of much larger extent north of Frazier Corners. Another extensive area is crossed by the Pacific Highway about five miles north of Redding. Smaller patches are present in the extreme southwest corner of the quadrangle.

Splendidly preserved fossils are found abundantly in the basal sandstones and conglomerates, in moderate quantities in the shales, and commonly in the upper sandstone and conglomerates. These fossils are of Upper Cretaceous age and belong to the middle part of the Chico epoch. Gastropods are most numerous, followed by pelecypods, and cephalopods. A large fauna is present, but unfortunately has not been described in detail.

The following is a section of the Chico formation measured along Dry Creek north of Frazier Corners in the south central part of the Redding quadrangle:

Red Bluff gravels (Pleistocene).
    Angular and erosional unconformity.   Feet

		Feet
7.	Gray shale, weathering lighter gray or brown on surface and containing abundant clay iron concretions. Thin interbeds of brown, arkosic sandstone.	40
6.	Gray shales similar to above with frequent interbedded brown, arkosic sandstones.	209
5.	Fine-grained, brown, arkosic sandstones with interbeds of gray shale.	56
4.	Massive brown, medium grained, arkosic sandstones with gravelly interbeds. Contain round, ferruginous concretions.	70
3.	Massive, brown, fine-grained arkosic sandstone. A few interbeds of shale and shaly sandstone some of which is lignitic.	183
2.	Massive, coarse-grained, brown arkosic sandstone with occasional shaly interbeds.	52
1.	Massive, coarse-grained brown, arkosic sandstones and interbedded conglomerates. At base, conglomerate composed of generally angular fragments of Triassic state.	40

    Angular and erosional unconformity.

Triassic
    Pit formation

Tertiary

EOCENE

Montgomery Creek formation

Along the southeast slopes of the Klamath Mountains in Kosk Creek and Pit River canyons, farther to the south on Montgomery Creek, and at a few other localities is a series of dominantly fluviatile brownish, arkosic sandstone, sandy shales, and conglomerates which Diller[14] in the Redding Folio called the "Ione" formation and dated Miocene on the basis of determination of fossil leaves made by Knowlton. These

14 Diller, J. S., *op. cit.*, p. 6.

strata have been recently examined by R. Dana Russell[45] who was associated with me in the geologic studies of the southern Klamath Mountains and will be described by him as the Montgomery Creek formation
in a report now in preparation. Determination by R. W. Chaney of a
flora discovered by Russell indicates a later Middle Eocene age for the
formation and Russell's field and laboratory studies prove that the
formation is not the Ione formation as Allen[46] has recently redefined
it along the eastern margin of the Great Valley.

The sediments appear to have come from the east and northeast
but the actual source is not known since most of the area covered by
these Eocene deposits was later flooded by Columbia-Cascade lavas and
pyroclastics. The detritus does not resemble bed-rock types of the
Klamath Mountain complex.

Only a limited area of this formation appears along the eastern
margin of the Redding quadrangle south of Little Cow Creek. Most
of the strata which Diller mapped as Ione belong to the Pliocene
Tuscan and Tehama formations.

Weaverville formation

Diller,[47] Hershey[48] and MacDonald[49] have described deposits of
"auriferous gravels" from a number of small basins in the western
Klamath Mountains; the principal areas are located at and near

Weaverville
and Shasta
Quadrangles

1. Trinity Center and Carrville
2. Lowden and Lewiston, along the Trinity River
3. Weaverville
4. Browns Mountain
5. Dutton Creek
6. Redding and Indian Creeks

Big Bar
Quadrangle

7. Hayfork
8. Hyampom

Under this heading of "auriferous or gold gravels," deposits of such
different ages and types are included that the term has no stratigraphic significance. The earliest or first cycle group, for example,
are the Cretaceous marine conglomerates which at certain localities
contain detrital gold. The second series includes fine-textured flood
plain sediments (sandstone, shaly sandstones, and sandy shales), lake
beds, lignitic shales and lignites, tuffs, and coarse stream gravels.
The fossil plants present toward the base of the series at various
localities are of Eocene age, but whether all of the beds belong to a
single sequence has not been determined. For this series of beds the
writer proposes the name Weaverville formation from the extensive
exposures near Weaverville, county seat of Trinity County. The gravels
most important as a source of gold are coarse channel deposits of

[45] Russell, R. Dana, Personal communication.
[46] Allen, V. T., The Ione Formation of California, Univ. Calif. Publ., Bull. Dept. Geol. Sci., vol. 18, pp. 337–448, 1929.
[47] Diller, J. S., Topographic Revolution in the Topography of the Pacific Coast: U. S. Geol. Surv., 14th Ann. Rept., p. 419, 1894. The Auriferous Gravels of the Trinity River Basin: U. S. Geol. Surv., Bull. 460, pp. 11-29, 1911. Auriferous Gravels in the Weaverville Quadrangle: U. S. Geol. Surv., Bull. 540, pp. 11-21, 1914.
[48] Hershey, O. H., The Sierran Valleys of the Klamath Region: Jour. Geol., vol. 11, p. 157, 1903.
[49] MacDonald, D. F., The Weaverville-Trinity Center Gold Gravels, Trinity County, California: U. S. Geol. Surv., Bull. 430, pp. 48-58, 1910.

Pleistocene age and similar sediments now being laid down by present streams.

The deposits of the second cycle have been preserved by the down-faulting of small blocks into the much more resistent bedrock of the region, and in these basins the soft, unconsolidated sediments were protected from rapid erosion.

The principal deposit extends from a short distance south of Weaverville for about twenty miles to the north northeast and has a width of one to three miles. Next in size are the exposures along Hay-fork stream at Hayfork and Hyampom in the Big Bar quadrangle, and on Indian and Redding creeks 10 miles southeast of Douglas City in the Weaverville quadrangle. The areas on Browns Mountain, along the highway two and one-half miles north of Lowden, along the Trinity River near Lowden, and on Dutton Creek southwest of Weaverville are much smaller. The deposits are composed chiefly of loose or slightly consolidated boulders, gravels, sands, and clays. Carbonaceous clays and sand and occasional seams of low grade lignite also are present. On Redding Creek and in the Hayfork and Hyampom basins, tuffs and tuffaceous sediments are present in considerable volume; tuffaceous sediments occur at Weaverville, Brown's Mountain, Dutton Creek, and the Lowden-Lewiston area. A section measured by Diller at this locality is given:

Redding Creek

```
    Partly cemented gravels_____400 feet
    Tuff and tuffaceous sediments_____250 feet
        In the upper part a 50-foot bed of gravel.
    Shale, thin beds of tuff and a bed of shaly coal 5 to 15 feet thick____250 feet
                Erosional and angular unconformity.
    Paskenta Lower Cretaceous and Subjacent Series.
```

Data regarding the other areas in which the Weaverville beds are found are contained in a report on The Auriferous Gravels of the Trinity River Basin, California, U. S. Geological Survey, Bulletin 470, pp. 11–29, 1911.

Plants found in the tuff and tuffaceous sediments on Redding Creek and near Hayfork and Hyampom were considered by the late F. H. Knowlton to be of Miocene age and to be equivalent to the fossil floras of the Sierra Nevadan auriferous gravels. According to Mr. Harry MacGinitie,[49a] who has worked on this flora under the direction of Dr. R. W. Chaney, the fossil plants are of Eocene age, and probably are to be correlated with the floras of the auriferous gravels of the Sierra Nevada.

Pliocene

Tuscan formation

One of the important superjacent formations which has been accumulated over the southeastern slopes of the Klamath Mountains and which extends for many miles down the east side of the Sacramento Valley is the huge pile of fragmental volcanics and interbedded sediments, which Diller[50] named the Tuscan formation. This sequence is composed largely of andesitic and basaltic fragments together with

---

[49a] MacGinitie, H., Personal communication.
[50] Diller, J. S., Lassen Peak Folio (No. 15), preliminary ed., U. S. Geol. Surv., 1892.

occasional dacitic tuffs. Agglomerates and tuffs are the chief constituents, while of subordinate importance are interbedded gravels, sands
and sandy clays, which are products resulting from the reworking of
the volcanic accumulations by streams. Diller believed the rocks to be
largely hypersthene andesite with minor quantities of hornblende
andesite and basalt, but Dr. C. A. Anderson of the University of California, who is making a detailed study of the Tuscan formation, states [51]
that basaltic fragments are present in much larger quantities than
Diller supposed.

The chief members of the Tuscan are thick mud-flows which spread
over a large area extending principally to the west of the present site
of Lassen Peak. To the east their distribution is obscured by a covering of later volcanics. In their advance to the north and west, the mud-
flow spread over an erosion surface cut in Eocene and Cretaceous
strata and finally reached and buried a part of the low foothills of the
southeastern Klamath Mountains. When this volcanic epoch had closed,
a great undulating table land had been developed to the southwest of
the Klamath Mountains. Subsequently streams deeply eroded the
plateau and stripped large volumes of the deposit from the underlying rocks. To the east, later volcanism buried this Tuscan landscape
beneath lava flows and the products of explosive eruptions. None the
less over large areas, tabular remnants of this mud-flow plateau are
exposed and are characterized by a bouldery surface apparently
developed by the removal of the finer detritus.

In Tehama County the Tuscan formation is at least 1000 feet thick
but along the southeastern slopes of the Klamath in the Redding quadrangle, the mud-flows thin out and generally less than 100 feet are preserved. The formation thickens rapidly from the west towards its
source. The nature and the exact location of the eruptive centers are
unknown, but they are believed to have been located in the general
vicinity of the present Lassen Peak.

Tehama formation

R. Dana Russell [52] has recently defined under the name of the
Tehama formation a series of beds abutting the southern Klamath foothills and extending far to the south in the Great Valley, which previously had not been recognized as a definite stratigraphic unit. This
formation consists of about ''2000 feet of massive, pale greenish gray
to pale buff silty sands, sandy silts and clays which are usually tuffaceous; intercalations of sand and gravel, often strongly cross bedded,
are present throughout. A massive coarse grained pumice tuff member
(called by Russell the Nomlaki tuff) is interbedded about 700 feet
above the base. This is of dacitic composition and consists of white
pumice fragments embedded in a medium to light gray matrix of glass.
and crystal shards. Most of the pumice fragments are less than three
inches in maximum diameter but they occasionally exceed one foot.
The upper six feet of the tuff often has a salmon pink color in marked
contrast to the predominant gray below. The tuff is variable in thickness; in Tehama County the maximum is fifty feet. On the eastern

[51] Anderson, C. A., Personal communication.
[52] Russell, R. D., and Vander Hoof, V. L., A Vertebrate Fauna from a New
Pliocene Formation in Northern California : Univ. Calif. Publ., Bull. Dept. Geol. Sci.,
vol. 20, pp. 11-17, 1931.

side of the Sacramento Valley a maximum of 300 feet was measured." [53] Fine exposures are located near Millville in the southeastern part of the Redding quadrangle where the rock was quarried for building stone. The Tehama beds are rather widely distributed in the southern part of the Redding quadrangle where they lie unconformably on the Chico Cretaceous. No exposures have been observed in the Weaverville or the mapped portion of the Red Bluff quadrangle.

In 1894 Diller [54] included in the Tuscan formation the Nomlaki dacite pumice tuff above referred to and thus greatly increased the area supposed to be covered by the Tuscan. The Nomlaki tuff is present near the base of the Tuscan formation at a number of localities on the eastern side of the Sacramento Valley. Russell [55] considers that it should be separated from the Tuscan because "of its highly distinctive appearance, dacitic composition in contrast with the andesitic composition of the Tuscan, much wider distribution, and value as a horizon marker." He therefore proposes for it the name Nomlaki tuff from the Nomlaki Indian Reservation in Tehama County where it is well exposed.

From various lines of evidence, Russell holds that the Tuscan and Tehama are contemporaneous formations. Stratigraphic data suggest that both probably are Pliocene and a vertebrate fauna collected near the base of the Tehama and above the Nomlaki tuff has proved to be upper Middle or Upper Pliocene in age. The two vertebrate horizons so far recognized are at 10 and 200 feet above the top of the Nomlaki tuff.

Evidently in Pliocene time eruptions of dacitic magma took place in the vicinity of the present Lassen Peak and the pumiceous fragments were swept far and wide over the Sacramento Valley by the fearful violence of the eruptions. The various deposits accumulated one over the other with no evidence of an important break between. The character of the eruptions suggests that probably the whole of the flood plain was not covered but considerable areas were so blanketed and a special type of initial volcanic surface was developed. This was volcanic in origin but different from the volcanic surface of the Tuscan which had started to form before and continued its construction long after the deposition of the Nomlaki tuff. Thus the andesite eruptions of the Pliocene in this region were much longer continued than the dacite, but apparently did not cover so large an area. After the cessation of the Nomlaki volcanism, andesite mudflows and sediments derived from them continued to form in the Tuscan area on the east side of the Great Valley, while on the floor of the Pliocene Great Valley flood plain sediments accumulated over the surface of the Nomlaki tuff to a depth of more than 1200 feet.

The Nomlaki outbursts represent a volcanic episode of comparatively short duration but of excessive violence. Beds of fine vitric tuff are present in the Tehama formation above the Nomlaki tuff but their compositions are not known. They show, however, that explosive eruptions recurred on a lesser scale in later Tehama time.

[53] Russell, R. D., Thesis summary. Univ. of Calif., 1931.
[54] Diller, J. S., Tertiary Revolution in the Topography of the Pacific Coast: U. S. Geol. Surv., 14th Ann. Rept., pt. II, p. 411, 1894.
[55] Russell, R. D., and Vander Hoof, V. L., loc. cit., p. 14.

Tertiary and Pleistocene lavas

In two areas, pyroxene-andesite and basalt flows of probable Pleistocene age are present:

1. At various places in the canyons of the Klamath and Siskiyou Mountains are remnants of narrow terraces cut by streams during a brief epoch of temporary base-leveling during the Cenozoic. Subsequent rejuvenation, apparently resulting from a regional elevation of the crust, has caused the streams to incise new gorges 50 to 100 feet below the terrace levels. Such a cut terrace is finely developed in the Little Sacramento Canyon; between the junction of Sugarloaf Creek and the Sacramento River, the terrace is capped in places by one, and in places by two flows of basalt which, according to Diller[56] "escaped from a vent surmounted by a cinder cone on the southern slope of Mt. Shasta, and, getting into the narrow valley of the river, followed it until the flow ceased and the lava congealed." Diller mentions only one flow, but below this is an older, less conspicuous, and more deeply weathered lava tongue which lies on the Sacramento terrace. After the second flow had solidified, the Sacramento ran over its surface for a time, and deposited upon it a considerable thickness of coarse gravels. Later, following the elevation previously referred to, the river cut the narrow gorge in which it now flows through the lavas and into the underlying bedrock.

Diller classed the flow described by him as andesite, but the plagioclase both of phenocrysts and groundmass is basic labradorite hence the rock is basalt. The flows are medium to dark gray in color and exhibit prominent porphyritic textures. Practically all of the phenocrysts are feldspars which show rude parallel arrangement. The plagioclase composes about half of the rock. Small grains of augite are just visible in hand specimens and are practically confined to the groundmass. Hypersthene and olivine are minor constituents. The ferromagnesian minerals compose about thirty per cent of the bulk of the rock. Brown glass, containing abundant small magnetite cubes, composes the residue.

2. Abutting the southeastern slopes of the Klamath Mountains and burying much of the post-Jurassic sediments and volcanics deposited over their lower foothills are flows of pyroxene andesite and basalt of Tertiary and Pleistocene ages. Near Kosk Creek and Pit River these lavas apparently belong to the Tertiary Cascade-Columbia sequence, but farther west the surface flows are more recent since locally they overlie the late Pliocene Tuscan formation. Unfortunately the history of this great lava covered region of northeastern California is little known. In the Tertiary and Pleistocene, great volumes of basic lava were erupted from fissures and from central vents and at certain centers through the entire region. Locally this activity has continued till very recent time; Lassen Peak is still active and it is impossible to say that the younger volcanoes and even some of the more eroded ones are extinct.

Splendid sections of the volcanics along their junction with the Klamath Mountains are to be found in the canyons of the Pit River and its tributaries east of Kosk Creek. These canyons are the deepest of any crossing the lavas and consequently should yield specially vital evidence in working out the history of the region. In the low hilly and

---

[56] Diller, J. S., Redding Folio, U. S. Geol. Surv., No. 138, p. 9, 1906.

mountainous country of the southeastern part of the Redding quadrangle, these lavas form caps on the broad divides between the streams.

The writer has not made petrographic studies of any of the probable Tertiary lavas but field examinations show that they are largely pyroxene andesite and basalts. Sediments, including diatomites, are interbedded with the flows. Little pyroclastic material is present in the sections which I have seen.

The later flows of probable Pleistocene age are chiefly pyroxene andesites which have come from fissures and vents in the Lassen volcanic area though not from Lassen Peak itself. Adjacent to the Klamath Mountains, no pyroclastics are present.

Quaternary

Pleistocene gravels

1. Red Bluff gravels of the Redding quadrangle

In the lowlands adjacent to the Klamath Mountains in the Redding quadrangle is a series of unconsolidated and poorly assorted gravels, sands, and occasional lenses of clay called the Red Bluff formation from fine exposures along the Sacramento River near the town of Red Bluff in Tehama County (Red Bluff quadrangle). The exposures are commonly reddish in color. These gravels extend in discontinuous areas up the principal canyons, notably that of the Sacramento. The deposit is composed principally of coarse detritus in which gravel, pebbles, and boulders predominate. The boulders, rounded or subangular, in many cases exceed a foot in diameter and some more or less angular boulders reach four and five feet. Numerous sandstone lenses and also occasional small lenses of shaly sand are present. The deposit is about 50 feet in thickness and overlies all of the earlier formations. It caps the old erosion surface extending from the foot of the Klamath Mountains southward to the margin of the Redding quadrangle. The deposit according to Diller is at least 100 feet thick along the Sacramento River near Redding but R. Dana Russell has shown that most of this cliff is cut in the Pliocene Tehama formation over which the Red Bluff forms a thin veneer. Remnants of gravels apparently once either continuous or nearly continuous with the Red Bluff of the plains are found at various localities in the Sacramento canyon. The texture of the deposit decreases in coarseness to the south and at the type locality near Red Bluff 30 miles south of Redding sand predominates and the beds of gravel and pebbles are of less significance.

2. Klamath gravels of the Weaverville quadrangle

At many places along the various canyons of the Weaverville quadrangle are deposits of boulders, pebbles, gravel, and sand similar in appearance to the Red Bluff gravels of the Redding quadrangle but generally coarser in texture. These deposits range from about 50 to more than 100 feet in thickness and have been considerably dissected by the streams flowing down through the canyons so that much of their original extent has been destroyed. Large volumes also have been removed in washing the gravels for gold. Apparently subsequent to the formation of the gravels the region suffered elevation which caused the rejuvenation of the streams and allowed them not only to cut through the gravel deposits but also for some distance into the underlying rock. The principal areas of these gravels are located (1) along Clear Creek northwest and southeast of Whiskeytown and on the tributary Whiskey Creek north of Whiskeytown; (2) about 1 mile west of the first occurrence on Clear Creek there is a long stretch of gravels running past Tower House and French Gulch to about a mile north of Cline Gulch, a tributary of Clear Creek; (3) an extensive deposit is located in the filled basin through which the Trinity River flows north of Bragdon Gulch and this deposit extends into the East Fork of the Trinity and also into Swift Creek on the west side; (4) on the East Fork of the Trinity River north of Minersville; (5) on the Stuart Fork of the Trinity more or less continuously from the mouth of Owen Creek to a point one-half mile from Minersville Post Office; (6) in the Big Meadows of the Stuart Fork between the mouth of Deer Creek and Devil's Canyon; (7) along the Trinity River from about 2 miles north of Lewiston past the mouth of Bear Gulch; (8) a small area just south of Lewiston; (9) along the Trinity River for about 3 miles east of Douglas City; (10) on Indian Creek from the mouth of Spring Gulch to the mouth of Mule Gulch; (11) along Redding Creek south of Clements Ranch. These deposits of gravel are of much economic importance because of their content of placer gold and during the whole history of mining in the Weaverville quadrangle have been extensively mined. These gravels were apparently deposited during the Pleistocene when the streams were supplied with an exceptional amount of coarse detritus. Possibly this occurred during the waning stages of the last glaciers which were present in the higher mountains of Trinity County and adjacent regions. The coarseness of the gravels suggests their deposition under flood conditions such as should have developed during the times of melting ice. The presence of boulders of exceptional size far down the courses of the streams is suggestive of ice rafting when the streams were swollen during times of thaw. The dissection of the deposits and the cutting of the streams into the bedrock indicate that a considerable time has elapsed since their deposition. The deposits themselves of course could be quickly channeled, but the bedrock, which has been trenched to a depth of 50 to 75 feet, is everywhere composed of very resistant rock. This dissection indicates a later re-elevation of the Klamath region on a moderate scale accompanied by important rejuvenation of the streams. The gravels overlie all other formations unconformably and when their deposition closed formed long narrow flats filling the bottoms of the submature canyons. Bones of Pleistocene animals have been found in the deposits in the Weaverville quadrangle.

These gravels are continuous with the Red Bluff deposits of the Redding and Red Bluff quadrangles. Since the gravels are so widespread through the canyons of the Klamath Mountains and since the deposits are so much coarser textured than the flood plain phase at the type locality near Red Bluff, the writer proposes for them the name Klamath gravels. The Red Bluff is an accumulation over a surface of low relief deposited by sluggish streams and should be designated as the Red Bluff phase. Upstream this phase grades into the coarser Klamath type near the city of Redding.

Recent alluvium

Unconformably overlying the Red Bluff and Tehama formations in the Redding quadrangle and also lying on all of the still older formations both superjacent and subjacent in all portions of the mapped area are deposits of gravels laid down by the present streams. In general these are characterized by coarse boulders many of which are 2 feet in diameter with which is mixed finer pebbles and coarse sand. The boulders include rock types characteristic of all earlier deposits. The principal deposits in the Redding quadrangle are along the Sacramento River, Clear Creek, and the various tributaries of Cow Creek; the texture of the deposits becomes finer along the lowland courses of the streams. These gravels contain a certain amount of gold grains which have been sought for by extensive placering. In the Weaverville and Red Bluff quadrangles because of the much greater diversity of relief the stream gravels are narrow deposits along the canyon bottoms. In the Weaverville quadrangle the main area of recent alluvium is in a basin through which the Trinity River runs north of Bragdon Gulch.

TRACING BURIED-RIVER CHANNEL DEPOSITS BY GEOMAGNETIC METHODS [1]

By ELMER W. ELLSWORTH [2]

Introduction

This article attempts to present a study of the application of geomagnetic methods of surveying to the development of certain types of placer ground. It is directed toward the problem of tracing the courses of relatively deeply buried, auriferous channel deposits, especially those of the Tertiary rivers in California. The application of the magnetic method to the development of similar deposits with a shallow depth of cover has been brought out in earlier reports,[3] and therefore will not be discussed here in detail. The writer has felt that if it could be adequately demonstrated that geomagnetic surveying may be used successfully in the mapping of these buried-channel deposits, this knowledge would be of interest and value to the placer mining industry as a whole. The results of the present investigation are presented as illustrating the usefulness of the geomagnetic method in this type of exploration.

Need for a proper field test of the application of geomagnetics to work of this kind was sensed in a statement made by Walter W. Bradley in an article entitled 'Renewed Activity in California Gold Mining,' appearing in the September, 1932 issue of Mining and Metallurgy: "The surficial placer accumulations of gold have now been fairly well skimmed off, but there are portions of buried, ancient river channels, as yet unworked, with gravel containing sufficient gold to be workable by drift mining; others, less concentrated, contain gold economically recoverable by hydraulic methods." Accordingly, in order to determine the possibilities and limitations involved in this use of the magnetic method of geophysical prospecting, the writer undertook the supervision of such a test, and during November, 1932, several weeks of field work were performed in Placer and Plumas counties, in the northern Sierra Nevada. Before proceeding it may be well to define the term 'geomagnetics' as it is used throughout this paper. Applied geomagnetics, which after all is the type of geomagnetics in which we are most interested, may be defined as the practical application of magnetic surveying to the solution of geologic problems.

The data here presented are based upon a geomagnetic survey on the Forest Hill Divide, which is situated in south-central Placer County, California. This area was selected as one suited to the making of such a test since the auriferous channels here are buried under several hundred feet of volcanic capping, and maps are available which accurately show their courses. This survey was concerned with the

---

[1] Published by permission of the Pacific Coast Geophysical Service of W. C. McBride, Inc.

[2] Consulting Geologist.

[3] (a) Arthur Gibson: Magnetometric Determinations Applied to Placer Mining. Eng. and Min. Journal, Vol. 114, No. 25: 1064. 1922.
    (b) K. C. Laylander: Magnetometric Surveying as an Aid in Exploring Placer Ground. Eng. and Min. Jour.-Press, Vol. 121:325–327. 1926.
    (c) C. A. Heiland and W. H. Courtier: Magnetometric Investigation of Gold Placer Deposits near Golden, Colorado. Geophysical Prospecting, 1929, A. I. M. E.: 364–384. 1929.

problem of tracing the gold-bearing Blue Gravel Channel, or 'blue lead,' which is concealed under 100 to 400 ft. of cap rock.

The geophysical work on the Forest Hill Divide described in this article is one of 30 similar investigations which have been conducted in California during the past two years by the Pacific Coast Geophysical Service of W. C. McBride, Inc. The Hotchkiss Superdip magnetometer has been employed in all of these surveys. This sensitive instrument measures relative variations in the total intensity of the earth's magnetic field. These variations from the normal, or magnetic anomalies, are caused by local geologic structures that aid or hinder, or otherwise affect the passage of the lines of force in the earth's normal magnetic field. For the survey at Forest Hill the sensitivity of the instrument was adjusted so that a one scale division change in its reading was equivalent to a variation in the local magnetic field of approximately 12.5 gammas.

The writer wishes to acknowledge the assistance given in the field by Mr. George Duffey, superintendent of the Mayflower Gravel Mines, and by Mr. Harold W. Kirchen of the Pacific Coast Geophysical Service.

Geologic Conditions at the Forest Hill Divide

At the Forest Hill Divide the lower portions of the ridges are composed of metamorphic and igneous rocks. These are overlain by a volcanic capping that has a maximum thickness of 300 to 400 ft. This capping includes volcanic agglomerate and other sedimentary strata. Most of the auriferous channel deposits, such as the Blue Gravel Channel, or 'blue lead,' which has an average width of about 1000 ft., occur between the bedrock and the overlying strata, while others are confined to the lower portions of the volcanic capping. The courses of these buried channels, which are all of Tertiary age, bear no relation to the present topography of the district, thus rendering their exploration a difficult problem. At Forest Hill the course of the Blue Gravel Channel has been very accurately determined, in the area covered by the present survey, as a result of extensive workings.

Details of the Recent Survey

Parallel lines of traverse were laid out with stations set every 50 ft. (every 100 ft. along traverse A-B). The locations of these lines of traverse are shown in Figure 1. Traverse line C-D follows, approximately, the surface trace of the Independent and Excelsior Slopes of the Blue Gravel Channel. This line of traverse coincides with the location of the geologic section of the Independent Tunnel and Excelsior Slope, which is shown upon the geologic map of the Forest Hill Divide accompanying the Tenth Annual Report of the State Mineralogist. The geologic section presented in Figure 2 has been reproduced from this original section, the location of the mine workings having been omitted for the sake of clarity. Traverse line A-B lies 1000 ft. to the northeast, while line E-F is 100 ft. southwest from line C-D.

A series of magnetometer readings were taken at each station, and their mean, together with the temperature and time, were recorded. By returning to a base station several times during the day a close check was maintained on the value of the diurnal variation in the earth's normal magnetic field. After completion of these traverses the instrument readings were corrected for temperature and diurnal variations in the usual manner. The corrected readings were then plotted, and the resulting 'magnetic profiles,' which reveal variations in the total intensity of the earth's magnetic field along the respective lines of traverse, are indicated in the accompanying Figures, 1 and 2.

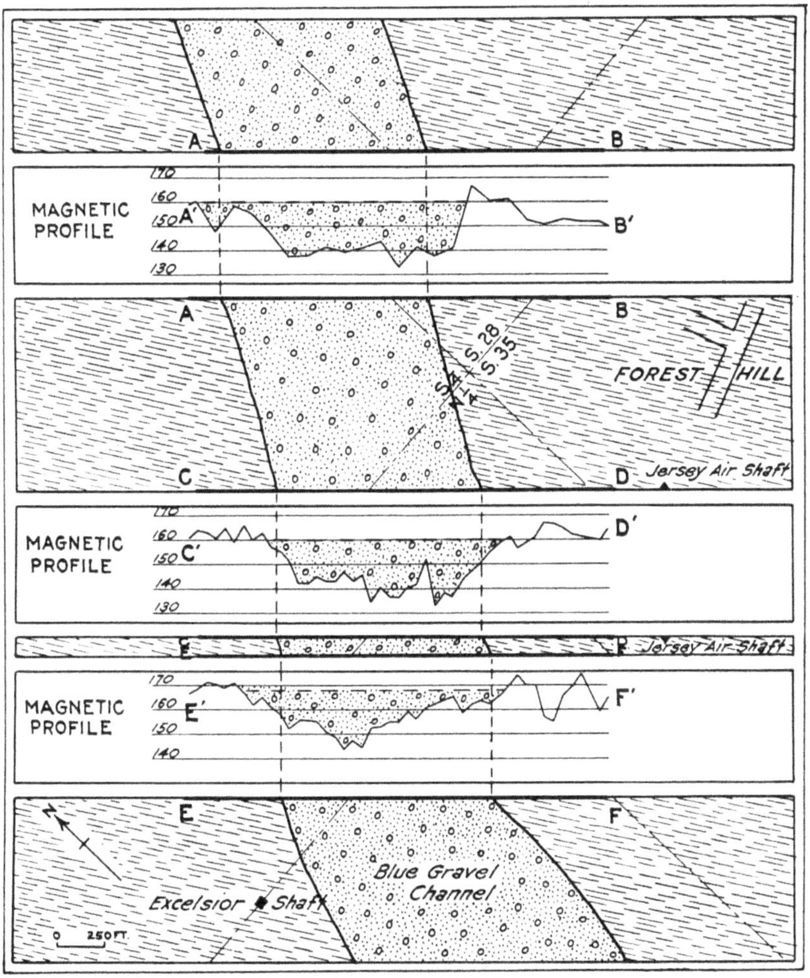

FIG. 1. Map of a portion of the Forest Hill Divide, Placer County, California, together with three parallel magnetic profiles across the Blue Gravel Channel. The course of the magnetic 'low' indicated by these three profiles is seen to coincide with that of this buried, auriferous channel. This geologic map has been taken from that which accompanies the Tenth Annual Report of the State Mineralogist.

Results of the Geomagnetic Survey on the Forest Hill Divide

The results of the survey are presented graphically in Figures 1 and 2. In Figure 2 is shown one of the magnetic profiles with a plan view and geologic section of the corresponding line of traverse, C-D.

The magnetic profile has one outstanding feature. This is the magnetic 'low,' or minimum in the total magnetic intensity, which occurs directly over the Blue Gravel Channel. This anomaly immediately suggests itself as being the magnetic reflection of this channel. Referring now to Figure 1, it is seen that all three of the magnetic profiles across this channel present anomalies which are not only similar, but which occur in each case directly above the channel. The course of the magnetic

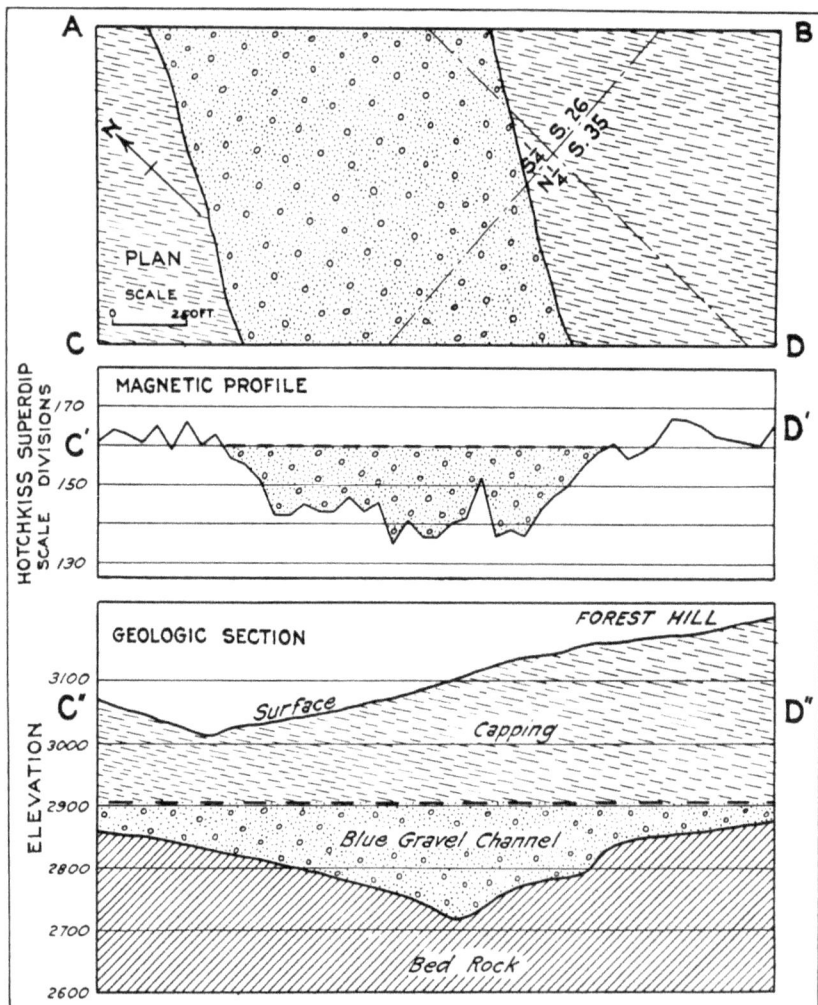

FIG. 2. Showing a magnetic profile across the auriferous Blue Gravel Channel at Forest Hill, and a plan view and geologic section of this ancient channel deposit. It is significant that the well-defined 'low' in the magnetic profile occurs directly above this river channel. The geologic data in this figure are taken from the map of the Forest Hill Divide accompanying the Tenth Annual Report of the State Mineralogist.

'low' thus mapped is found to coincide with that of the buried Blue Gravel Channel.

Interpretation of Results

A basic principle controlling the usefulness of magnetic surveying in this type of prospecting has been well stated by C. A. Heiland: "For the application of magnetic methods to the exploration of placer ground success can be predicted only if the placer carries *less* magnetite or *more* magnetite than the country rock."[4] A study of the local stratigraphy in connection with the magnetic data presented above leads to the conclusion that these channel gravels and the cap rock are less magnetic (have a lower magnetic permeability) than the underlying metamorphic and igneous rocks. Thus the magnetic profiles reveal a decreased intensity, or magnetic 'low,' over those sections of a line of traverse in which the bedrock is the most deeply buried, as in the center of a large channel. The rather uniform results obtained above the Blue Gravel Channel on the Forest Hill Divide appear to confirm this interpretation. Stated otherwise, the magnetic profiles obtained at Forest Hill appear to reflect the profile of the bedrock itself. In 1926 Heiland concluded[5] that somewhat similar magnetic minima were found by K. C. Laylander[6] over less-deeply buried, gold-bearing channel deposits at Keithley and Cedar Creeks, British Columbia.

The commonly accepted belief that concealed channel deposits will always be detected in magnetic profiles by rather pronounced magnetic maxima, or 'highs,' due to concentrations of magnetic minerals in such deposits, is seen to have important exceptions. A proper interpretation of the data obtained with the magnetometer can be made only after a thorough study has been made of all geologic factors concerned. Where channel deposits have but a slight depth of cover, and where the bedrock is virtually nonmagnetic, their presence may be detected in the magnetic profiles by magnetic 'highs.' The results of the present survey tend to show that in detecting the presence of relatively-deeply buried channel deposits, the possible presence of local concentrations of magnetic minerals in the channel itself is a factor which may be less important than others, such as the magnetic permeabilities of both the cap rock and the bedrock, as compared with that of the channel deposit as a whole. The present investigation thus indicates that it may be possible to map buried channels by magnetic methods irrespective of their content of magnetic minerals, the surveys in such cases furnishing data on the character of the bedrock topography. Arthur Gibson likewise states that he has "*   *   * obtained good results by magnetometric surveys of placer deposits, and where there was no evidence of the presence of any magnetic (attractile) minerals."[7] In such cases the magnetic method is less direct, but as illustrated by the survey on the Forest Hill Divide, may prove to be highly effective in obtaining the desired results.

Conclusions relative to the general use of geomagnetic methods in the exploration for buried channel deposits

Depending upon local geologic conditions, the application of geomagnetic methods will vary:

---

[4] C. A. Heiland: Prospecting with the Magnetometer. Eng. and Min. Jour. Vol. 122 : 59. 1926.
[5] C. A. Heiland: *op. cit.*, 61.
[6] K. C. Laylander: *op. cit.*
[7] Arthur Gibson: The Magnetometer as an Aid in the Development of Mines. Min. and Scientific Press, Vol. 123 : 437. 1921.

(1) Where there is no appreciable cover of later deposits, a geomagnetic survey can be relied upon to locate and outline those parts of river deposits in which black sand concentrations occur. While these concentrations generally will contain the highest values to be found in the deposit, the indicated presence of such a concentration is not in itself a guaranty that any gold values will be found. This is another basic principle that merits a clear understanding.

(2) Where channel deposits, such as the Tertiary river deposits of the northern Sierra Nevada, are buried under a capping of several hundred feet in thickness, the value of the geomagnetic method, as demonstrated at Forest Hill, lies not so much in the location of local areas of possible concentration as in the determination of the course of the buried channel as a whole—detecting its presence, tracing its course, and indicating its approximate width. While occasional mineral concentrations in the concealed channel deposit may be reflected in the magnetic profiles, their depth of burial will in many cases be too great to warrant their exploration except by operations designed to work the channel as a whole.

It should be pointed out that in some areas the distribution of the magnetic minerals in the cap rock is so irregular as to render this method impracticable in the tracing of underlying channels. South Table Mountain, near Oroville, California, was found by Arthur Gibson to be an example of such an area. For this and similar reasons it is very desirable that a preliminary test survey be made in the field in order to determine the application of the geomagnetic method to the area of immediate interest. Many geophysical exploration companies, including the one sponsoring the present investigation, are not only willing to conduct these test surveys at no cost to the client—except for actual expenses in the field—but ordinarily require that such preliminary surveys be made, for the protection of both their client and themselves. If this survey produces negative results, the advantage of finding this out before, rather than after money has been spent on a thorough examination of the properties in question is quite apparent.

General considerations concerning the application of geophysical methods

While in this article mention has been made of only the geomagnetic method, and while the writer is of the opinion that this is the most rapid and least expensive of all the geophysical methods of exploration, it should be pointed out that electrical methods have also been successfully applied to the problem of locating auriferous channel deposits, as recently reported by International Geophysics.[5] The electrical surveys were followed by magnetic surveys, which lent partial confirmation to the results obtained by the electrical method. The following discussion is intended to apply to other geophysical methods as well as the magnetic.

There are two distinct types of problems in which geophysical surveying can render service in the development of placer ground:

(1) In already partially developed ground it may aid in tracing auriferous channel deposits from known workings into virgin territory. As an example of a discovery of this kind, made as the result of a geo-

---

[5] J. J. Jakosky and C. H. Wilson: Use of Geophysics in Placer Mining. Min. Journal, Dec. 1932.

magnetic survey, there is the report of the results of the survey of the Maxine deep gravel mine in Plumas County, California, made by Arthur Gibson in 1921. "The magnetic survey disclosed an ancient river channel containing heavy mineral concentrations, south of and close to the Maxine shaft, between 450–500 feet wide, extending from east to west, with a slight southward trend; * * *. A tunnel was driven south from the old workings, and the main channel was tapped. The values as well as the average depth proved to be a great improvement over the old workings." [9]

(2) In virgin territory geophysical surveying can be employed to advantage in the attempt to outline the most likely areas of mineral concentration. Such surveys lead the way for exploration by the customary methods of test-pitting, drilling, etc. An outstanding success in this application of geomagnetic methods was reported at Keithley Creek, British Columbia, by K. C. Laylander. In planning the development of this new placer ground Laylander states that "Pipe lines, power house, and equipment were distributed with these (geomagnetic) data as a sole guide. The placer ground has been worked this year and conforms exactly to that indicated by the magnetometer." [10] In this type of exploration problem the value of a geophysical survey lies not only in detecting the presence of any channel deposits, but in saving time and money in the subsequent testing of the most promising areas. In this connection it should be emphasized, for the protection of those most interested in the success of such adventures, that any geophysical survey is but the first step, essential as it may be, in the exploration work, and that it should always be immediately followed by the usual, and incidentally more expensive methods of mineral exploration. A favorable geophysical report should not be taken alone as proof of the value of a property, but rather as a favorable indication that deserves further investigation. Geophysical prospecting is not a substitute for the old-established methods of exploration, and the realization of this fundamental principle by prospective employers of geophysical services will not only place these methods in their proper light, but will lead the way toward a mutual understanding of the problems involved, and a fuller cooperation between the geophysicist, on the one hand, and the mining engineer or geologist, on the other.

It is perhaps well to recognize that as the surficial placer deposits become more and more worked over, and as more and more attention is directed toward the development of buried river channels, new methods of exploration must be found that are suited to the tasks at hand. The writer believes that properly supervised geophysical surveying is very much a part of the new technique in the successful exploration for those concealed deposits of gold-bearing gravel that today constitute a significant source of new gold in California.

[9] Arthur Gibson (1922), op. cit., 1069.
[10] K. C. Laylander, op. cit., 326.